SpringerBriefs in Applied Sciences and Technology

Manufacturing and Surface Engineering

Series Editor

Joao Paulo Davim, Department of Mechanical Engineering, University of Aveiro, Aveiro, Portugal

This series fosters information exchange and discussion on all aspects of manufacturing and surface engineering for modern industry. This series focuses on manufacturing with emphasis in machining and forming technologies, including traditional machining (turning, milling, drilling, etc.), non-traditional machining (EDM, USM, LAM, etc.), abrasive machining, hard part machining, high speed machining, high efficiency machining, micromachining, internet-based machining, metal casting, joining, powder metallurgy, extrusion, forging, rolling, drawing, sheet metal forming, microforming, hydroforming, thermoforming, incremental forming, plastics/composites processing, ceramic processing, hybrid processes (thermal, plasma, chemical and electrical energy assisted methods), etc. The manufacturability of all materials will be considered, including metals, polymers, ceramics, composites, biomaterials, nanomaterials, etc. The series covers the full range of surface engineering aspects such as surface metrology, surface integrity, contact mechanics, friction and wear, lubrication and lubricants, coatings an surface treatments, multiscale tribology including biomedical systems and manufacturing processes. Moreover, the series covers the computational methods and optimization techniques applied in manufacturing and surface engineering. Contributions to this book series are welcome on all subjects of manufacturing and surface engineering. Especially welcome are books that pioneer new research directions, raise new questions and new possibilities, or examine old problems from a new angle. To submit a proposal or request further information, please contact Dr. Mayra Castro, Publishing Editor Applied Sciences, via mayra.castro@springer.com or Professor J. Paulo Davim, Book Series Editor, via pdavim@ua.pt.

Diego Carou

Introduction to Generative Design for Aerospace Applications

From the Creative Process to Manufacturing

Springer

Diego Carou
Departamento de Deseño na Enxeñaría,
Campus As Lagoas
Universidade de Vigo
Ourense, Spain

ISSN 2191-530X ISSN 2191-5318 (electronic)
SpringerBriefs in Applied Sciences and Technology
ISSN 2365-8223 ISSN 2365-8231 (electronic)
Manufacturing and Surface Engineering
ISBN 978-3-031-89052-9 ISBN 978-3-031-89053-6 (eBook)
https://doi.org/10.1007/978-3-031-89053-6

© The Author(s), under exclusive license to Springer Nature Switzerland AG 2025

This work is subject to copyright. All rights are solely and exclusively licensed by the Publisher, whether the whole or part of the material is concerned, specifically the rights of translation, reprinting, reuse of illustrations, recitation, broadcasting, reproduction on microfilms or in any other physical way, and transmission or information storage and retrieval, electronic adaptation, computer software, or by similar or dissimilar methodology now known or hereafter developed.
The use of general descriptive names, registered names, trademarks, service marks, etc. in this publication does not imply, even in the absence of a specific statement, that such names are exempt from the relevant protective laws and regulations and therefore free for general use.
The publisher, the authors and the editors are safe to assume that the advice and information in this book are believed to be true and accurate at the date of publication. Neither the publisher nor the authors or the editors give a warranty, expressed or implied, with respect to the material contained herein or for any errors or omissions that may have been made. The publisher remains neutral with regard to jurisdictional claims in published maps and institutional affiliations.

This Springer imprint is published by the registered company Springer Nature Switzerland AG
The registered company address is: Gewerbestrasse 11, 6330 Cham, Switzerland

If disposing of this product, please recycle the paper.

Preface

For centuries, designers relied on pencils and paper to create drawings of parts and assemblies. With the advent of digital technologies in the late 20th century, technology has helped to reshape the design landscape. Computers, equipped with powerful simulation software, have become dominant tools for sketching, drawing, and 3D modeling. This shift has allowed designers and engineers to optimize their products without physical prototypes or initial testing.

In response to the highly demanding global market, which now requires customization, high flexibility, and price reduction, the role of design, materials science, manufacturing, and production has become increasingly critical. While there have been significant advances and innovations in mature processes, researchers and industry are now focusing on developing disruptive technologies that can deliver substantial cost savings and increase flexibility. New simulation tools like generative design, coupled with suitable manufacturing methods such as additive manufacturing, are paving the way for a transformative era in manufacturing, meeting the market's evolving needs.

Structural optimization methods, such as shape, size, and topology optimization, have been developed and increasingly applied in the last decades. Notably, significant progress in computer science since the 1980s has helped the development of new and improved methods for structural optimization. These methods offered designers new ways to optimize their designs, pursuing objectives such as weight reduction. In recent years, generative design tools have garnered interest for their unique ability to explore the design space and develop different results that consider mass, stress, and displacement, as well as materials and manufacturing. Generative design takes advantage of recent technological advancements such as the processing power of modern computers, cloud computing, and, in some cases, big data and data analytics. The tool allows designers to create geometries that can resemble biological systems, resulting in significant mass savings. This is of great importance in sectors such as aerospace and, more broadly, for the transportation sector, as sustainable production continues to gain prominence in the global economy.

Nowadays, tools driven by artificial intelligence, such as those used to generate text or images, are becoming extremely popular for various applications. In the industrial field, generative design is a technology that allows the creation of new designs almost from scratch with the aid of computers. The topic is also getting attention from researchers, software developers, and design and manufacturing companies that use and promote generative design. This book aims to contribute to this novel design methodology by providing readers with a comprehensive understanding of its principles and applications.

Notably, the aerospace sector is highly innovative and develops numerous applications involving new technologies such as additive manufacturing. In engineering, particularly aerospace engineering, fostering connections between academia and industry is essential. Thus, it is crucial to help transfer knowledge from companies to universities efficiently.

The book *Aerospace and Digitalization. A Transformation Through Key Industry 4.0 Technologies* (Carou 2021) presents state-of-the-art Industry 4.0 enabling technologies in the aerospace sector. This book follows the approach of its predecessor and aims to provide a background for an advanced tool such as generative design. Moreover, it aligns with the objectives of the *Grupo de Innovación Docente "Empresa e Innovación"*, which seek to connect academia and industry through innovation and technology.

This publication introduces generative design and its applications in the aerospace sector to students, researchers, professionals, and the general public. The objective is to offer a blueprint of the topic using a general perspective that starts with creativity and ends in manufacturing. Chapter 1 opens the reader's eyes to the book's contents. Then, Chapter 2 explores the concept of creativity by highlighting some milestones. Chapter 3 introduces the framework of sustainable production and focuses on manufacturing and assembly. Chapter 4 digs deeper into the recent technological advances that allow moving to a more "machine-driven" designing process. In recent years, there has been a transition from drawings on paper to computer-aided drawings that has ended in the digitalization of the product development process. Moreover, artificial intelligence is already a key element in the future that will help refine optimization methods to a great extent. Structural optimization, reviewed in Chapter 5, has a long tradition. This chapter presents the foundations of generative design, including size and shape optimization. Special attention is given to topology optimization, which is the closest to generative design. Chapter 6 presents the novel generative design methodology to help designers develop new ideas and concepts to work on. Moreover, the chapter shows a wide range of applications for the aerospace sector. Finally, Chapter 7 presents the book's final observations and conclusions.

Baiona, Spain
January 2025

Diego Carou

Reference

Diego Carou. 2021. *Aerospace and digitalization. A transformation through key industry 4.0 technologies*, ed. J.P. Davim. Springer Nature.

Competing Interests The author has no competing interests to declare that are relevant to the content of this manuscript.

Contents

1 **Introduction** 1
   References 3
2 **The Creative Trail** 5
   References 10
3 **Sustainable Production** 13
   3.1 Manufacturing 14
       3.1.1 Machining 17
       3.1.2 Additive Manufacturing 19
   3.2 Assembly 20
   References 22
4 **All About Technology** 25
   4.1 The Foundations 25
   4.2 From Sketches to Digital Products 27
   4.3 Artificial Intelligence 31
   References 33
5 **Structural Optimization** 37
   5.1 Size, Shape, and Topology Optimization 38
   5.2 Optimization Methods 40
   5.3 Topology Optimization Examples 41
   References 42
6 **Generative Design and Its Applications** 45
   6.1 The Concept 46
   6.2 Exploring Differences Between Generative Design
       and Topology Optimization 49
   6.3 The Workflow 50
   6.4 The Biological Inspiration 52

	6.5	Generative Design Examples	55
	6.5.1	Aircraft Bearing Bracket	56
	6.5.2	Aircraft Engine Loading Brackets	57
	6.5.3	Bionic Partition	60
	6.5.4	Landing Gear of an Ultralight Single-Seater Aircraft	60
	6.5.5	GE Bracket	62
	6.5.6	Drone Frames	63
	6.5.7	Wing Ribs of a Remote-Controlled Plane	64
	6.5.8	Flapping Wings	65
	References		66
7	**Concluding Remarks**		71
	Reference		73

Chapter 1
Introduction

Abstract In recent years, machines have started to perform tasks once considered exclusive to humans. The rise of artificial intelligence has opened a debate on whether it can be adequate to perform routine tasks and tasks deemed creative. In this regard, milestones such as IBM's Deep Blue defeating a chess champion and Watson competing on Jeopardy received broad interest from the media and the public. The rapid advance of technology and artificial intelligence is creating new opportunities for innovation in design. New generative design tools are being developed and improved, and while still requiring human oversight, they can create designs in which creativity is enhanced.

The rapid development of computing, big data, and advanced data analytics in recent years has enabled machines to perform tasks that were hardly expected in the past. Artificial intelligence (AI) has emerged as a central element in specific sectors of the economy and is increasingly influencing human activities, finding applications beyond the industrial economy.

Artificial intelligence technology can efficiently perform routine tasks, but is also learning to develop more complex activities over time. Artificial intelligence has received growing interest from academia, mainstream media, and the general public. Each day, newspapers and TV news provide novel examples of artificial intelligence capabilities to solve different tasks. Solving these complex tasks has always been related to humans; thus, in a sense, artificial intelligence is challenging humans' purpose.

Mass media highlighted the artificial intelligence potential through milestones such as Deep Blue, a chess-playing computer developed by IBM (Campbell et al. 2002), defeated a world chess champion such as Garry Kasparov in 1997, demonstrating artificial intelligence's potential in specialized tasks (Sinha and Marimuthu 2024) or how the IBM's Watson artificial intelligence contested in a popular trivia TV show, Jeopardy, in 2011 (van der Maas et al. 2020). News also presented the first auction of an AI-made artwork at Christie's in 2018, selling the painting for $432,500 (Magni et al. 2024). These milestones are being succeeded more and more frequently, showing the potential of artificial intelligence to a large share of the human population.

The same idea that one may have of a machine has been profoundly transformed in recent years. Thus, large, heavy-duty machines of dubious aesthetic value have evolved into refined devices of smaller size, lighter weight, and more attractive designs. This refinement can be seen in the design of the latest laptops, smartphones, robots, and home appliances.

Recent advances in artificial intelligence enable us to envision a future in which artificial intelligence can directly compete with humans in activities that seem exclusive to humans. Throughout history, humans have already given up in favor of technological progress. For example, math tasks have been increasingly performed by calculators and spreadsheets. Similarly, the Internet has become an incredible source of information globally. Thus, proficiencies such as computing or retaining information have become less relevant over time.

Humans have shaped the world according to their needs, thanks to creativity. This has helped them discover the laws of nature and use existing resources in multiple ways to create transformative inventions. Human beings can solve problems by integrating different kinds of knowledge acquired through their experiences (Chulvi et al. 2022). This creativity is innate in people (Amabile 1996) and a sign of human identity. As Simon Colton and Geraint Wiggins pointed out, "As a society, we are jealous of our creativity" (Colton and Wiggins 2012). At least, this can be the case so far. But at this point, a new path is now opening up for things to change.

Today, artificial intelligence is beginning to create and be creative. The mechanisms of this creativity are different from those of humans. But, in certain respects, artificial intelligence can achieve results that humans cannot achieve unaided. In this context, humans may want to safeguard the niche of emotional and social competence and creativity, aspects that presently cannot be "coded", as questioned by Neubauer (2021). No matter how artificial intelligence evolves, there is a tension between the appreciation of the creativity of AI-made artifacts and the fear and anxiety of being replaced by generative AI (Magni et al. 2024). According to Neubauer (2021): "In the future, AI systems might, for instance, challenge human cognitive capacity". However, whether this challenge is on the horizon or humans are already grappling with its implications is a matter for discussion.

As machines face tasks traditionally associated with humans, relevant aspects of humans and society come into discussion. This is not only true in academic discourse; many material and personal resources must be deployed to solve associated problems such as privacy, consent, accountability, and the potential for bias (Beltracchi 2024). Thus, aspects such as machine ethics are fundamental in this new era.

Generative AI[1] has become a powerful tool for numerous material science and drug design applications. In addition, according to the technology research and consulting firm Gartner, generative AI shows potential in developing parts for automotive and vehicle manufacturing, as well as in architecture and engineering. This approach is globally recognized as "generative design".

---

[1] https://www.gartner.com/en/articles/beyond-chatgpt-the-future-of-generative-ai-for-enterprises.

Generative design has emerged as a tool for driving design to the next stage. With computers and powerful software aided by artificial intelligence, users are given improved capabilities to explore the design space more effectively, obtaining a wide range of optimized solutions that are nearly ready for manufacturing. Generative design provides designers with complex designs that would be difficult to obtain without this tool, but the process still requires human oversight and input.

References

Amabile, T.M. 1996. *Creativity in context: Update to the social psychology of creativity*. Boulder, CO: Westview Press.

Beltracchi, Carlo. 2024. Resilience and the metaverse: a toolkit approach. In: *Coding architecture. Designing toolkits, workflows, industry*, ed. Pierpaolo Ruttico. Springer.

Campbell, Murray, A. Joseph Hoane Jr., and Feng-hsiung Hsu. 2002. Deep Blue. *Artificial Intelligence* 134: 57–83.

Chulvi, Vicente, Marta Royo,·María-Jesús Agost,·Francisco Felip, and Carlos García-García. 2022. How the type of methodology used, when working in a natural environment, affects the designer's creativity. *Research in Engineering Design* 33: 231–248.

Colton, Simon, and Geraint A. Wiggins. 2012. Computational creativity: The final frontier? *Frontiers in Artificial Intelligence and Applications* 242: 21–26.

Magni, Federico, Jiyoung Park, and Melody Manchi Chao. 2024. Humans as creativity gatekeepers: Are we biased against AI. *Journal of Business and Psychology* 39: 643–656.

Neubauer, Aljoscha C. 2021. The future of intelligence research in the coming age of artificial intelligence—With a special consideration of the philosophical movements of trans- and posthumanism. *Intelligence* 87: 101563.

Sinha, Piyush Kumar, and Marimuthu R. 2024. Conglomeration of deep neural network and quantum learning for object detection: status quo review. *Knowledge-Based Systems* 288: 111480.

van der Maas, Han L.J., Lukas Snoek, and Claire E. Stevenson. 2021. How much intelligence is there in artificial intelligence? A 2020 update. *Intelligence* 87: 101548.

Chapter 2
The Creative Trail

Abstract Although many definitions of creativity have been proposed, no universally accepted one exists. Traditionally, creativity is understood as the generation of ideas that are novel and useful. In this sense, it is possible to recognize creativity in various fields, such as architecture and industrial design. In the end, creativity is a key driver of innovation. This chapter traces the evolution of creativity from the early Oldowan stone tools to the Fiskars tools with their characteristic orange sleeves. The drivers of creativity and innovation, namely aesthetics, ergonomics, functionality, and identity, are introduced through examples that illustrate how these drivers have influenced the development of design throughout history.

Creativity is widely recognized across many disciplines, from the design of vehicles for streets and sports competitions to the worlds of fashion and international cuisine. The leading creative minds receive great media and social attention in their respective fields. In many cases, the fame of these creative minds is accompanied by biographies, documentaries, exhibitions, and dozens of articles in magazines and newspapers. For instance, the industrial designer Dieter Rams, the chef Ferran Adrià, the fashion designer Jean-Paul Gaultier, the designer Jonathan Paul Ive (Jony Ive), and the architect Zaha Hadid are widely recognized for their groundbreaking contributions to creativity and innovation.

The concept of creativity has become enormously popular and is widely used in all aspects of life. But what does it represent? Although everyone understands the idea, defining it and reaching a shared definition is challenging. Creativity cannot be measured or ranked objectively; it depends significantly on the observer.

Conventionally, creativity consists of two parts. One is novelty, and the other is usefulness, effectiveness, or value (Sternberg 2022). In this sense, creativity is "the generation of ideas that are both novel and appropriate or useful" (Amabile 1996; Oldham and Cummings 1996; Shalley et al. 2015).[1] The idea of novelty and usefulness depends on time and place. Thus, it is not just the observer but also the

---

[1] There are multiple definitions of creativity. Readers can find a list with some definitions in Research in Engineering Design 32, 2021, 289–307.

© The Author(s), under exclusive license to Springer Nature Switzerland AG 2025
D. Carou, *Introduction to Generative Design for Aerospace Applications*,
Manufacturing and Surface Engineering, https://doi.org/10.1007/978-3-031-89053-6_2

time and place of the observation that are essential elements in defining creativity. As Sternberg (2022) points out, current definitions of creativity are generally relativistic.

Creativity is a key driver of innovation. Thus, it is possible to view a broader process composed of creativity and innovation. Creativity concerns idea generation, whereas innovation concerns implementing creative ideas (Shalley et al. 2015). Ultimately, ideas need to be materialized to demonstrate their usefulness.

The ability of human beings to create can be traced back hundreds of years. Written records are the primary source of study on humans' creative capacity. The Italian artist Leonardo da Vinci was one of the greatest minds ever. The drawings and explanations of da Vinci in the Forster Codices of the late 15th and early sixteenth centuries are some of the most outstanding examples, which can be found the Victoria and Albert Museum in the United Kingdom.[2] Da Vinci made many discoveries by mixing his knowledge and interests across multiple areas. His inventions merged anatomy, science, and technology into futuristic designs that were ahead of their time. Visionaries have helped advance science, pointing to new directions by exploring novel concepts. The French writer Jules Verne could also imagine technological advances long before they were realized. *From the Earth to the Moon*, published in 1865, is a fictional book in which Verne described a lunar journey, paving the way for space exploration.

While da Vinci and Verne lived centuries apart, most of their work has reached present times, and their historical context is easily understood. When compared to the origins of writing (i.e., around five thousand years ago), they could be considered almost contemporaries. But what has happened to creativity since the emergence of human beings, long before writing existed? Where and when can we trace the earliest signs of creativity?

The absence of written records makes it extremely difficult to identify the motivations behind creative activities. This was the case with the Egyptian civilization's way of life, customs, and beliefs. This imposing society left an immense legacy and remained largely unknown for hundreds of years. The discovery of the Rosetta Stone in Egypt in 1799 was crucial for interpreting hieroglyphs, completed after Jean-François Champollion's translation in the 1820s. Understanding the hieroglyphs allowed for understanding this civilization, particularly some drivers that moved Egyptians to create such impressive architecture and artwork.

When written records are absent, objects and the remains of structures provide insight into the cultures of antiquity. Historians and archaeologists are responsible for shedding light on the past. In this regard, the role of archaeological excavations is crucial. Thanks to the excavations, it is possible to know and classify the use of materials over time once the findings are dated. This is reflected in the classification of the three ages, based on the use of stone, bronze, and iron materials.

Most of the antique objects found in the excavations share a common feature: they were created based on functionality. Archaeological museums around the world are an excellent example of this. These museums gather a large number of artifacts of different natures. Their creators had to carry out creative activities to respond to

---

[2] www.vam.ac.uk.

a necessity for their everyday activities, such as hunting, eating, and dressing. To this end, these early designers had to identify the need and envisage a solution. They must have a good command of the available materials and their processing methods, weight, and geometries to obtain objects that guarantee the expected requirements. As Ross and Glăveanu (2025) pointed out, there is a distinction between craft and creativity, with craftsmanship being more of a case of repetition and creativity a more disruptive process. Although early designs were primarily focused on utility, the ideation behind these objects was a game changer, marking the emergence of creative thinking. Therefore, creativity can be reasonably associated with these early objects, as they were not merely products of repetitive craftsmanship.

Tools are an excellent example of objects designed with a clear function. Throughout centuries, cutting tools like knives have evolved, often adapted for combat and warfare across various historical settings. The case of the Oldowan stone tools, found in the Olduvai Gorge, Tanzania, is a primitive demonstration of creativity in which natural materials were adapted for practical purposes. These tools arise when extracting sharp blades from the core of a corner (Stout 2016). In this case, creativity is not related to the generation of a complex and novel design but to the identification of a natural element with the right properties and geometry to address a specific need (e.g., for hunting and butchering animals). These findings date back to the Lower Paleolithic (2.6 million to 200,000 years ago).

The creative process throughout history was not solely guided by functionality. As soon as it was possible to have suitable methods, another type of design approach emerged. Thus, aesthetics emerged as an objective in new designs. These new objects were designed for aesthetics to symbolize power, tribal affiliation, or social status. Designers also had personal goals trying to create aesthetically pleasing objects that showcased their art and ability to increase relevance, distinguish themselves from their peers, and even achieve lasting significance. One clear example of this is jewelry.

Undoubtedly, when thinking of the quest for aesthetics, one of the leading human activities is art. The Renaissance is one of the most outstanding historical periods in terms of artistic dynamism. Florence is one of the fundamental places to understand this period and its achievements extending into modern times. Aesthetics is the main engine of the novel artistically designed creations, mainly in architecture, sculpture, and painting. Indeed, not all this development can be based on aesthetics. Still, power or the need to showcase power and wealth were important motivations for the primary beneficiaries of these projects, coming from the government and the church.

For millennia, many human societies centered their daily activities on agriculture, livestock, and fishing alongside trade. As cities developed and grew, other activities, such as construction, increased their economic share. However, a rapid and profound transformation occurred not until the First Industrial Revolution. While the design process is an evolution of the existing one, its economic and social importance increased substantially during these times.

The first significant technological developments associated with the First Industrial Revolution were oriented toward functionality and problem-solving, although the aesthetics of some of these solutions were recognized later. An example is the boring machine created by John Wilkinson and the screw-cutting lathe created by

Henry Maudslay (Sharma 2022). During this period, machining and the pursuit of improved precision were crucial to the industry. An example is the development of the Watt steam engine was significantly enhanced by the precision of cylinders produced using John Wilkinson's boring machine (Winchester 2021).

Functionality continued to be a paradigm in industrial design during the Second Industrial Revolution. One of the main symbols of this period is the car. The Ford Motor Company, founded by Henry Ford, is a reference company that helps one understand the origins of the automotive industry. Ford developed the assembly line for mass production,[3] making cars accessible and affordable to the general public. The objective was to sell an affordable car with basic features for people (Alizon et al. 2009). Between 1908 and 1927, the company produced more than 15 million units of the Model T (Fiell and Fiell 2021).

As the industry advanced, the industrial design evolved to a new dimension. Functionality remained the basis for creating new products. However, aesthetics, guided by the pursuit of modernity and a constant demand for innovation, soon emerged as a defining characteristic of industrial product identity. In the globalized economy that emerged in the second half of the twentieth century, global competition increased notably. During this period, the design of products and everything associated with sales strategies, such as advertising or packaging, gained great relevance, with functionality and aesthetics being core characteristics and success factors of products, as highlighted by Han et al. (2021).

Fiskars is a Finnish company founded in 1649. However, it was in the 1960s[4] that the company introduced new lines of products with striking designs, thanks to their characteristic orange sleeves, which managed to set the focus of attention on the company using a distinctive marketing claim. Fiskars was also working on ergonomics, which gained importance by that time (Fiell and Fiell 2021). These motivations in creating new products are evident in the ergonomically oriented Swedish company Ergonomi Design Gruppen, which created products for customers as diverse as Sandvik or Scandinavian Airlines System (SAS) (Fiell and Fiell 2021).

Another interesting example from the twentieth century is the case of Gillette. King Camp Gillette developed a revolutionary safety razor with handles and interchangeable blades, patented in 1904 (Fig. 2.1a). Despite this notable innovation, the American company continued developing new designs for over a century, always maintaining the basic concept of the interchangeable blade, from its rudimentary first patented model to its Blue Blade, Mach 3, and Fusion ProGlide Power models (Fig. 2.1b). In these models, functionality merges with ergonomics and aesthetics (Fiell and Fiell 2021).

The industry's power in shaping society in the last two centuries is unquestionable. People's everyday lives would not be the same without the thousands of industrial devices and various types of equipment that have been created. Industrial design

---

[3] Oldsmobile, a company founded by Ransom E. Olds make significant contributions to mass production of cars.

[4] https://fiskarsmuseum.fi/en/explore-learn/the-digital-museum/modern-innovations-of-the-1960s-and-1970s/.

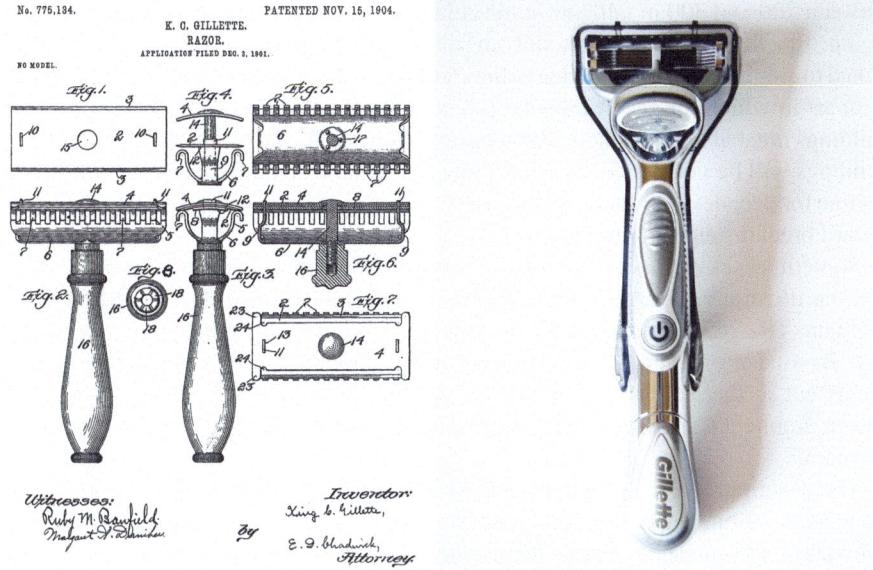

Fig. 2.1 Gillete: (left) Patent 1904,[5] (right) Gillette Fusion ProGlide Power[6]

played an essential role in this transformation, but human creativity extends beyond industry, and valuable examples can be found in activities such as architecture. Functionality and aesthetics have been constant over the centuries in any design process, with the architecture of spaces where people live, work, learn, and relax serving as a prime example.

Architecture offers impressions heavily engraved in the collective imagination. It can transform the way cities and even countries are perceived by society. In this regard, skylines often serve as symbols representing a city and are proof of dynamism, economic power, and modernity. As a case in point, to think of the City of New York is to think of its skyscrapers. The Empire State and the Chrysler Building are good examples in New York. It is interesting to see how, despite New York being a city built in height, these buildings are iconic for many reasons and exert a specific magnetic power on people. This is evident in the number of visitors who visit the city each year to see them.[7]

The philosophy of skyscrapers has spread worldwide since the beginning of the twentieth century (Ahlfeldt and Barr 2022). Many countries currently have buildings

---

[5] By K. C. Gillette—US patent 775134, p. 1, Public Domain, https://commons.wikimedia.org/w/index.php?curid=14859097.

[6] By Carnby—Own work, CC BY-SA 3.0, https://commons.wikimedia.org/w/index.php?curid=28719915.

[7] 68 million visitors in 2022. Source: https://www.statista.com/statistics/1447123/total-visitors-nyc-us/.

between 200 and 300 m tall, and some cities also have buildings surpassing 500 m.[8] Again, functionality and aesthetics are objectives to pursue in these new designs linked to commercial interests due to the significant investment required. Architecture is an art in which the result must be functional and structurally stable. In addition, buildings must also be aesthetically pleasing (Song et al. 2016). In many cases, these buildings will be visible from virtually any point in the city and will define the city's skyline for decades. This is why the variety of designs is so large, especially the trend toward breakthrough, futuristic styles.

Sometimes, skyscraper designs seek a symbiosis with society and the city's identity. An illustrative example is the Bitexco Financial Tower in Ho Chi Minh City, Vietnam. The 262-m-high and 68-story building has quickly become an icon of the city. Beyond the beauty of the building itself, which is always subjective, a fact stands out in the design process. In the creative process of the Bitexco Financial Tower, architect Carlos Zapata draws inspiration from the lotus flower, a symbol of Vietnam.

The design process also allows the creation of personal identities through brand value association. There are numerous examples of this approach to creativity and innovation. For instance, Apple, the technology brand based in Cupertino, is one of the most valuable and influential companies worldwide and is widely associated with creativity and innovation, partly due to strategic decisions such as offering exclusive, high-priced products. Apple's products have been regarded as a symbol of status. Undoubtedly, multiple Apple customers have sought association with these values, in a way, sharing Apple's values of creativity, innovation, and exclusivity.

References

Ahlfeldt, Gabriel M., and Jason Barr. 2022. The economics of skyscrapers: A synthesis. *Journal of Urban Economics* 129: 103419.

Alizon, Fabrice, Steven B. Shooter, and Timothy W. Simpson. 2009. Henry Ford and the Model T: Lessons for product platforming and mass customization. *Design Studies* 30 (5): 588–605.

Amabile, T.M. 1996. *Creativity in context: Update to the social psychology of creativity.* Boulder, CO: Westview Press.

Fiell, Charlotte, and Peter Fiell. 2021. *El diseño industrial.* Taschen: Bibliotheca Universalis.

Han, Ji, Hannah Forbes, and Dirk Schaefer. 2021. An exploration of how creativity, functionality, and aesthetics are related in design. *Research in Engineering Design* 32: 289–307.

Luz Castro Pena, M., Adrián Carballal, Nereida Rodríguez-Fernández, Iria Santos, and Juan Romero. 2021. Artificial intelligence applied to conceptual design. A review of its use in architecture. *Automation in Construction* 124: 103550.

Oldham, Greg R., and Anne Cummings. 1996. Employee creativity: Personal and contextual factors at work. *Academy of Management Journal* 39 (3): 607–634.

Ross, Wendy, and Vlad Glăveanu. 2025. The constraints of habit: craft, repetition, and creativity. *Phenomenology and the Cognitive Sciences* 24: 251–271.

---

[8] Currently, the Burj Khalifa skyscraper (828 m), in Dubai, is the tallest building in the world. Source: https://www.ctbuh.org/.

References

Shalley, Christina E., Michael A. Hitt, and Jing Zhou. 2015. *The Oxford handbook of creativity, innovation, and entrepreneurship*. Oxford University Press.
Sharma, Prince. 2022. A brief account of man, material and manufacturing: On the timeline. *Materials Today: Proceedings* 66: 3572–3577.
Song, Hwayeon, Jamshid Ghaboussi, and Tae-Hyun Kwon. 2016. Architectural design of apartment buildings using the implicit redundant representation genetic algorithm. *Automation in Construction* 72: 166–173.
Sternberg, Robert J. 2022. Missing links: What is missing from definitions of creativity? *Journal of Creativity* 32: 100021.
Stout, Dietrich. 2016. ¿Cómo nos cambió la fabricación de herramientas? *Investigación y Ciencia*, 28–35.
Winchester, Simon. 2021. *Los perfeccionistas. Cómo la precisión creó el mundo moderno*. Turner Publicaciones.

Chapter 3
Sustainable Production

Abstract In the last few years, there has been a growing concern about the evolution of the environment and, particularly, the impact that human activities have on it. One key challenge is to attain sustainable production processes, especially in industries such as aerospace and automotive. Generative design is a tool that can help optimize material usage. In transportation, this is key as it reduces fuel consumption and, thus, environmental impact. Generative design is linked to a manufacturing or production process, depending on the objective of creating a part or an assembly. Therefore, in the chapter, some of the most critical manufacturing processes relevant to generative design, such as additive manufacturing, machining, and assembly, are briefly presented.

As the world population rapidly grew from approximately four to eight billion people over just 50 years,[1] there was a growing concern about the pressure human activity puts on the environment, particularly about how goods and services are produced. There has been little concern about production processes for decades after the beginning of the 1st Industrial Revolution. Still, in recent years, the industry has been required to limit the environmental impact of its activities. To do that, several established strategies under the sustainable production concept exist, such as product design based on circularity and redesign of supply chains. For instance, the industry's influence on environmental impact is notable when considering global greenhouse gas emissions (Hanenkamp and Zipse 2023; Cooper et al. 2024).

Sustainability awareness is mainly associated with growing concerns about global climate change. In the past, global initiatives have been launched to address this issue. In 1972, the United Nations hosted the Conference on the Human Environment in Stockholm. One of the results was the creation of the United Nations Environment Programme (UNEP). In 1988, the UNEP, in collaboration with the World Meteorological Organization (WMO), formed the Intergovernmental Panel on Climate

---

[1] In 48 years, from 1975 to 2023, the world population doubled. Source: https://www.statista.com/statistics/1303469/time-taken-for-global-pop-double/.

Change (IPCC)[2] (Beckmann et al. 2023). Nowadays, the *Transforming our World: The 2030 Agenda for Sustainable Development* (United Nations 2015) is the framework that guides the road toward sustainability (Khajuria et al. 2022). This agenda includes 17 Sustainable Development Goals (SDGs) (Andreadakis 2023). The 12th goal is "responsible production and consumption" (United Nations 2015).

Some objectives to pursue in sustainable production include minimizing transport, embedding functions, reducing waste, improving resource efficiency, and designing lightweight products (Dani et al. 2020). Sectors such as automotive and aerospace recognize the need to reduce weight and, thus, decrease greenhouse gas emissions. Weight reduction also offers significant economic benefits, but this strategy must ensure safety (Pascual et al. 2023).

Sustainable production involves various ideas and strategies (e.g., energy and circularity) that are not the specific focus of this work. Moreover, although materials[3] are a fundamental input of the generative design and critical for designing sustainable production processes, this topic will not be specifically covered. For the purpose of this book, manufacturing and assembly are the crucial areas related to the outcomes of the generative design process. Generative design can optimize raw material usage and help redefine the way assemblies are made. Thus, assemblies can be created using fewer parts, which can enhance durability.

Many manufacturing processes that could benefit from generative design are available. However, efforts must be made to select the most important ones to avoid overextending this section. Machining is still a relevant manufacturing process in aerospace, and optimizing part geometries for machining can benefit companies and the environment. Additive manufacturing is proving to be a technology capable of creating parts with significantly reduced weight compared to conventional ones. Moreover, assembly processes are being rethought, sometimes eliminating them, based on the capabilities of additive manufacturing. The following sections will briefly introduce machining, additive manufacturing, and assembly.

3.1 Manufacturing

Although other results are possible (e.g., innovation in organization, procedures), the primary outcome of a creative process in the industrial field is typically a part or a product. The manufacturing process is critical for obtaining parts according to the specified requirements. The associated creative process must focus on manufacturing, as it imposes constraints that limit the designer's freedom. Notably, the designer must attend to the materials, machines, tools, jigs, and fixtures used for manufacturing.

---

[2] IPCC monitors climate change. In its *Climate Change 2023: Synthesis Report*, the panel alerts that global greenhouse gas emissions have continued to increase due to unsustainable energy use, land use and land-use change, lifestyles, and patterns of consumption and production. Source: https://www.ipcc.ch/report/ar6/syr/downloads/report/IPCC_AR6_SYR_SPM.pdf.

[3] Professor Michael F. Ashby published a book on the topic entitled *Materials and Sustainable Development* (2016).

3.1 Manufacturing

It is essential to contextualize and understand the evolution of manufacturing from the twentieth century to the present day. Manufacturing systems have primarily transitioned from mass production to the smart factory concept in the twenty-first century. Table 3.1 summarizes the central manufacturing systems, their applications, the new technology in each period, and their main features (Zhu 2022). It is important to note that manufacturing is highly related to digital tools that offer flexibility and efficiency today.

Although older systems that may still coexist with modern ones, the leading manufacturing systems are currently defined as computer-integrated manufacturing and cyber-physical production systems (or cyber-physical systems).

The computer-integrated manufacturing (CIM) system involves extensive computer utilization. It is related to the initial development of computers in the

Table 3.1 Evolution of manufacturing systems

Period	Name	Application	New technology	Feature
1940–1950	Automatic production line	Mass production	Relay program control and machine tool combination	Rigid and high productivity
1950–1980	CNC machine tool machining center	Multi-species production of single or more	Numerical control (NC) and Computer numerical control (CNC)	Flexible process focus
1970–1980	Flexible manufacturing systems	Multi-species production of small batches. Mass production	CAD, robot, group technology, DNC, and automation	The ideal combination of flexibility and efficiency
1980–present	Computer-integrated manufacturing system	Factory of automation of design, manufacturing, and economic management	CAD/CAM/CAPP, production management and scheduling, information technology, and simulation technology	Fully automated, optimized, intelligent, and distributed communication network
2014–present	Cyber-physical production system	Smart factory. Digital workshop	Digital twin, dynamic scheduling, cloud computing, and big data analytics	Adaptation to changing environments and smart decision

Based on Zhu (2022)

1970s. The integration of computers into manufacturing facilitates the seamless coordination of previously separate tasks such as engineering design, factory operations, and business functions (Groover 2016). The cyber-physical system (CPS) concept, coined by the US National Science Foundation (Maamar et al. 2023), represents a fusion of "cyber" with "physical" worlds (Alcácer and Cruz-Machado 2019; Tao et al. 2019). The cyber-physical system integrates different fields of knowledge, such as computer science, information and communication technology (ICT), and electrical and mechanical engineering (Darwish and Hassanien 2018; Wang et al. 2015). The CPS is a step forward from the CIM concept in which digitalization is a key differentiator.

Some driving forces for the evolution of the systems are the inclusion of automation and the search for more flexible manufacturing systems. The current possibility of capturing and processing large amounts of data enables advancement toward the cyber-physical production system, refining digital models based on the actual outputs of the processes and improving them using advanced digital models. This strategy can be performed using a continuous improvement approach, with manufacturing technologies as central elements.

Various manufacturing routes exist, including casting, powder metallurgy, plastic injection, and metal shaping. These processes are often combined under a suitable sequence that takes advantage of the main benefits of the different processes. Among these, subtractive processes have received significant attention since the First Industrial Revolution. Notable advances have made machining a reliable process capable of producing complex parts with improved accuracy and surface quality. Moreover, since the 1980s, significant investments have been made in developing diverse additive manufacturing technologies, which are widely recognized as a disruptive technology. Because the process has already provided examples of use in different sectors with essential benefits, more investments are being made to move the technology forward. In this regard, additive manufacturing and machining are particularly well-suited for producing various parts for the aerospace industry. Recent developments in manufacturing technologies allow to obtain increasingly complex parts. Thus, it is possible to identify a paradigm shift from "what can be manufactured" to "what can be imagined and validated", as Essink et al. (2017) stated.

The comparison of additive and subtractive methods is of particular interest. When dealing with material scrap, subtractive methods tend to generate large amounts of chips that may be recovered and recycled. In contrast, most additive manufacturing methods only require using the material needed to build the part (and, when required, the material for supports). In terms of complexity, despite advances in cutting tools and modern CNC centers that can help generate more complex geometries in machining, factors such as clamping and physical limitations regarding movements still limit the creation of geometries that the layer-by-layer process can produce, such as highly complex inner geometries.

3.1.1 Machining

Machining is a subtractive process in which a cutting tool shapes a workpiece by removing material in the form of chips. There are several types of processes and machines to choose from, depending on the shape of the workpiece. For instance, cylindrical and prismatic workpieces are preferably manufactured using turning and milling operations, respectively. Cutting tools can be either single-edge or multi-edge cutting tools. Moreover, the movements of the tools can be arranged in different ways (i.e., tool or workpiece rotation, workpiece displacement, and/or rotation) under multi-axis configurations. Thus, machining is a process that offers flexibility and helps engineers manufacture a wide range of parts, but a good understanding of the process is essential when designing.

Significant innovations in cutting tool materials have been made in the last decades. For instance, synthetic and natural diamonds have been employed as cutting tools in ultra-precision machining (Zhang et al. 2024). The development of polycrystalline diamond (PCD), high-temperature high-pressure (HTHP) diamonds, and the chemical vapor deposition (CVD) route for coating applications are some significant milestones that have greatly improved machining.

5-axis machines have increased their share in manufacturing because they allow for creating more complex geometries and save time due to the reduced number of setups. However, a significant drawback is the machine tool accuracy, especially in terms of geometric accuracy (Jiang et al. 2022). In Fig. 3.1, a machine tool with a 5-axis configuration is visible. The machine offers two additional rotational moves in addition to the three conventional displacements.

There are specific requirements in aerospace due to the complex types of parts to be manufactured. For instance, there is a need to manufacture large parts. XiangFei et al. (2022) presented some examples of manufacturing complex geometries using multi-axis machining. For instance, machining a large outlet guide vane decreased the deformation from 3 to 0.62 mm in rough milling and attained an accuracy of the finished parts of 0.2 mm. Another example relates to machining an integral stator blade ring using a multi-axis strategy with only one clamping. Thus, it was possible to avoid alignment errors. This is critical because the necessary precision was ± 0.065 mm, given the large integral stator blade ring's diameter (over 1 meter).

Robots have been integral to industrial operations since the second half of the twentieth century. In the context of the Fourth Industrial Revolution, automated and collaborative robots are increasing their role in manufacturing. In this sense, robotic machining is a strategy being analyzed to replace humans and manufacture parts that may be inefficient to manufacture in conventional machine tools due to costs, shape, or for improving integration among processes.

For large complex surfaces, robotic machining offers advantages such as ample operating space, good dexterity, and easy-to-realize parallel machining (XiangFei et al. 2022). However, robotic machining is not especially suited for applications that require high material removal rates (Ji and Wang 2019) and presents other challenges, such as low accuracy and low stiffness of the robots as compared to the conventional

Fig. 3.1 5-axis machine configuration (Caputi and Russo 2021)

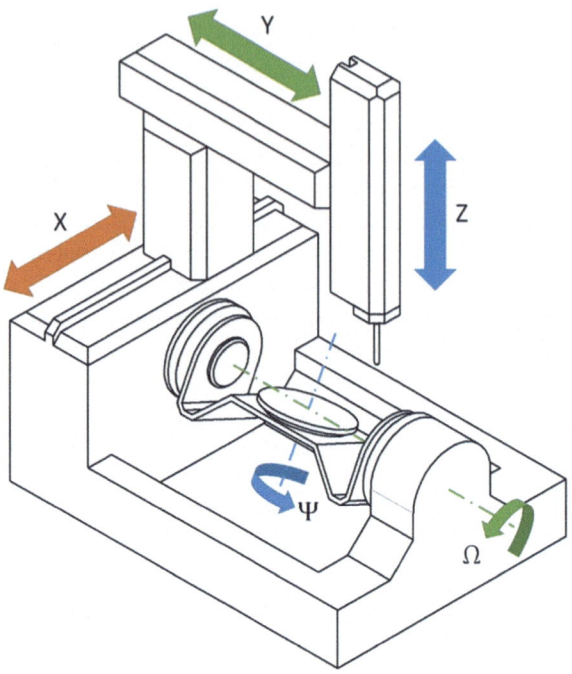

machine tools (XiangFei et al. 2022). In this sense, the main applications of robotic machining include processes with lower machining loads, such as polishing and grinding (Kim et al. 2019). There is already a long tradition of research in robotic machining. For instance, in 1987, Appleton and Williams presented a study in which a robot performed several machining applications, including drilling, grinding, and deburring (Ji and Wang 2019). Figure 3.2 illustrates an experimental setup for robotic machining and measuring.

One of the most interesting advantages of robots is their mobility. Thus, larger spaces can be reached. Several examples of the use of mobile robotic systems can be found in the literature, such as the KUKA Moiros mobile robot developed by the Fraunhofer Institute, the mobile robot developed by Université de Nantes, and the mobile Hexapod robotic machining unit proposed by Tunc and Shaw (Bo et al. 2019). The Fraunhofer Institute has proposed an example of this strategy in aerospace to mill the integral wall panel of the A320 vertical tail (XiangFei et al. 2022).

The need to create micro-manufactured parts has driven micromachining, a process in which the parts are in the micrometer range (Masuzawa 2000). Mechanical micromachining has proven to be an adequate process to obtain microscale features from various materials for the aerospace, energy, and medical sectors (Shekhar et al. 2023). For instance, micro-milling can provide high accuracy, small machining size, and complex machining surface (Ding et al. 2023), making microtool manufacturing especially critical. Non-conventional machining processes (e.g., electrical discharge

3.1 Manufacturing

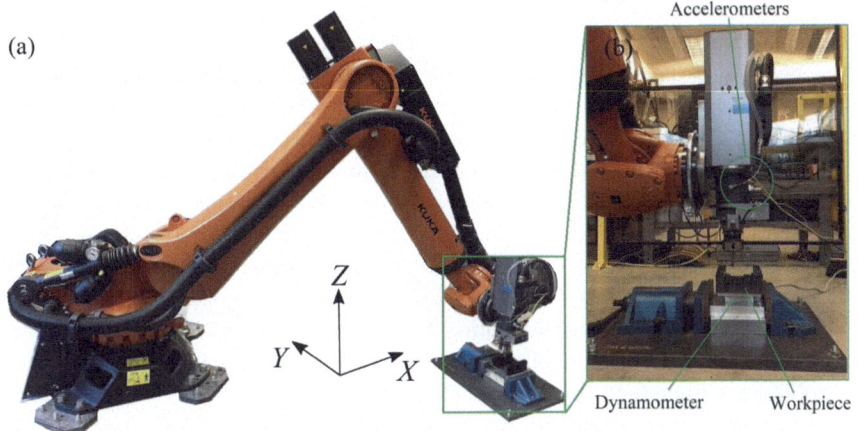

Fig. 3.2 Robotic machining: (left) KUKA KR90 R3100 robotic milling system, (right) Experimental measuring setup (Mohammadi and Ahmadi 2023)

machining, focused ion beam machining, and electrochemical machining) are suitable for this purpose due to their lower cutting forces, non-contact machining, and the small size of the machinery (Kumar et al. 2021).

Size reduction is a growing priority for physics, chemistry, electronics, materials, and biology applications. As the demand for smaller features grew, microfeatures were no longer sufficient, leading to the development of nanofabrication techniques in the late twentieth century. The invention of the Scanning Tunneling Microscope (STM) by Gerd Binnig and Heinrich Rohrer in 1981 (Elzein 2024) was worth noting. Several processes are available in nanofabrication, including tip-based nanomachining, which utilizes an atomic force microscope (Kong et al. 2020). The process is a high-resolution and low-cost nanomanufacturing process. However, it is critical to guarantee the tool's sharpness to reach the quality of the machined patterns (Cheng et al. 2022).

3.1.2 Additive Manufacturing

Additive manufacturing is a technology that offers advantages, such as manufacturing complex geometries without limitations that result in optimized parts regarding weight. The workflow starts with a 3D (CAD) model designed from sketches or scans of existing objects (Nyamekye et al. 2021). This manufacturing strategy is especially suited to take advantage of design strategies such as topology optimization, generative design, bio-inspired design, and lattice structure filling (Wang et al. 2023). Additive manufacturing has opened the way for advanced concepts to be developed. Thus, there is abundant research on novel topics such as multi-materials (du Plessis et al. 2019; Nazir et al. 2023), lattices structures (Wang et al. 2021), embedded intelligence, structural colors, and metamaterials (Xiong et al. 2022).

Notable efforts have been made to develop additive manufacturing, and today, the concept has resulted in several processes. The ISO/ASTM 52900 standard (ISO/ASTM 2022) helps categorize additive manufacturing into seven process categories: binder jetting, directed energy deposition, material extrusion, material jetting, powder bed fusion, sheet lamination, and vat photopolymerization. Each process has unique characteristics that make it especially suited for different applications. These processes offer designers and engineers alternatives for processing materials such as plastics, ceramics, and metals. Despite the availability of some machines, limitations, such as low resolution and high printing times, must be considered for manufacturing large-scale parts (Bacciaglia et al. 2022). Some notable advances have been accomplished in manufacturing large-scale parts, such as the printing of rockets, as illustrated by the cooperation between Relativity Space and NASA.[4]

Additive manufacturing has been used in aerospace to manufacture different prototypes and parts. For example, 1000 different parts were manufactured for the Airbus A350 XWB. A famous application example is the fuel nozzle designed by GE Aviation for the CFM LEAP turbofan engine. Several other examples from companies such as Ariane Group, GKN Aerospace, Honeywell Aerospace, Lockheed Martin, and Northrop Grumman can be found in the literature. The technology is also used in space projects by the National Aeronautics and Space Administration (NASA) and the fabrication of satellites, such as CubeSats (Carou 2021).

Stavropoulos (2023) studied the feasibility of a nozzle of a rocket engine, which is 1000 mm in diameter and 900 mm in height. The parts were to be manufactured in Inconel 718, and because productivity is critical, directed energy deposition (DED) processes were selected. The machine had to be modified to achieve the minimum thickness of the part's features. The process allowed for the obtaining of a part that fulfilled the requirements. The case study helps one realize that additive manufacturing can be very profitable for an aerospace application, offering advantages such as improved product performance, technical advantages, and fewer production steps, eliminating various operating, maintenance, and consumable costs. Moreover, as the technology is still developing and increasing its role in the industry, the authors claimed that the profit margin will grow steadily in the coming years.

3.2 Assembly

Modern supply chains are highly complex and distributed worldwide. For example, in aerospace, supply chains may include thousands of suppliers organized into three levels (tiers). The leading aircraft manufacturers, Airbus and Boeing, have supply chains of more than 12,000 firms (Carou 2021).

---

[4] www.relativityspace.com.

3.2 Assembly

Although some products might seem simple, one's initial perception can change when counting the number of individual components involved. Thinking in terms of parts is a good starting point for understanding the importance of assembly; for instance, the Boeing 747–800 range comprises six million individual parts (Budd and Ison 2020), with about half of them being fasteners (McCarthy and McCarthy 2015). Thus, it is essential to consider the operations required to assemble and secure these parts.

In aerospace, it is not just the number of parts that makes assembly a complex task; it is also their large dimensions and the strict tolerances to be met. In the case of drilling, circularity geometric constraints of ± 0.03 mm are needed, and the operation center must comply with a position tolerance of ± 0.50 mm. Moreover, the requirement for absolute positional accuracy may reach values up to ± 0.20 mm (da Silva Santos et al. 2022).

The assembly of an aircraft largely determines the manufacturing cost, product quality, and delivery cycle. Assembling the thin-walled aircraft parts is traditionally done using mechanical connections (rivets and bolts) and bonding. Drilling and installing fastener holes are usually done manually. Recently, to avoid the low machining accuracy, low-quality consistency, high defect probability, high labor intensity, and health hazards of manual work, automated drilling and fastening systems are being researched and deployed (Mei et al. 2023). However, automation is still challenging (da Silva Santos et al. 2022). Other methods for joining parts include fusion joining (low heat input laser beam welding) and solid-state joining operations (friction stir welding) (Tsirkas 2018). For fiber-reinforced polymer matrix composite structures, welding has been used for fixed leading edges, rudders, and elevators on Airbus A380 and Gulfstream G650 airplanes (Ma et al. 2024).

As the complexity of the products increased, new design approaches were needed. In this sense, the Design for Manufacturing and Assembly (DFMA) was popularized in the 1980s. It has many applications in the automobile, aerospace, and other manufacturing industries (Naiju 2021). This methodology was devised as a tool that connects design with later stages of the process, aiming to create a design that helps increase efficiency throughout the entire process. DFMA helps reduce manufacturing time and cost (Pudi et al. 2023).

Curran et al. (2004) presented a case study of a pressure box. The assembly is a pressure-box that functions as one of two cavities located between the floor beams in the pressurized mid-fuselage section of a regional jet aircraft. The original design included sheet metal fabrication with joining using fasteners, and the proposed alternative solution, machining. The Design for Assembly (DFA) solution allowed reducing parts from 29 to 1, the number of fasteners from 346 to 124, and the assembly time from 20 to 3.3 h. The authors also discussed the Design for Manufacturing (DFM) application for the firewall bulkhead of a tail-cone on a Lear 45 business jet, reporting savings in raw material from 143 to 96 kg and machine time from 138 to 90 h. DFMA was also used in Bell Helicopters, as reported by Naiju (2021).

The emergence of additive manufacturing defies traditional design processes, as it reduces the required parts. The conventional DFMA methodology still needs to be adapted for additive manufacturing, which has given rise to a new design approach: the design for additive manufacturing (DfAM) (Gromat et al. 2023).

References

Alcácer, V., and V. Cruz-Machado. 2019. Scanning the industry 4.0: A literature review on technologies for manufacturing systems. *Engineering Science and Technology, an International Journal* 22 (3): 899–919.

Andreadakis, Stelios. 2023. The future of corporate purpose: Merging and balancing social, environmental and economic considerations. In: *Corporate government for climate transition*, ed. Carolina Machado, and João Paulo Davim. Springer.

Appleton, E., and D.J. Williams. 1987. *Industrial robot applications*. New York: Halsted Press.

Bacciaglia, Antonio, Alessandro Ceruti, and Alfredo Liverani. 2022. Towards large parts manufacturing in additive technologies for aerospace and automotive applications. *Procedia Computer Science* 200: 1113–1124.

Beckmann, Markus, Gregor Zöttl, Veronika Grimm, Thomas Becker, Markus Schober, and Oliver Zipse. 2023. Setting the course for net zero. Translating climate science into political and corporate targets. In: *Road to net zero. Strategic pathways for sustainability-driven business transformation*, ed. Oliver Zipse, Joachim Hornegger, Thomas Becker, Markus Beckmann, Michael Bengsch, Irene Feige, and Markus Schober. Springer.

Bo, Tao, Zhao XingWei, and Ding Han. 2019. Mobile-robotic machining for large complex components: A review study. *Science China Technological Sciences* 62 (8): 1388–1400.

Budd, Lucy, and Stephen Ison. 2020. Air transport management. In: *An international perspective*, 2nd edn. New York: Routledge.

Caputi, Antonio, and Davide Russo. 2021. The optimization of the control logic of a redundant six axis milling machine. *Journal of Intelligent Manufacturing* 32: 1441–1453.

Carou, Diego. 2021. *Aerospace and digitalization. A transformation through key industry 4.0 technologies*, ed. J. P. Davim. Springer Nature.

Cheng, Fei, Shi-Chen. Zhai, and Jingyan Dong. 2022. Investigation of Gaussian mixture clustering model for online diagnosis of tip-wear in nanomachining. *Journal of Manufacturing Processes* 77: 114–124.

Cooper, Samuel J. G., Stephen R. Allen, Ahmed Gailani, Jonathan B. Norman, Anne Owen, John Barrett, and Peter Taylor. 2024. Meeting the costs of decarbonising industry—The potential effects on prices and competitiveness (a case study of the UK). *Energy Policy* 184: 113904.

Curran, R., S. Raghunathan, and M. Price. 2004. Review of aerospace engineering cost modelling: The genetic causal approach. *Progress in Aerospace Sciences* 40: 487–534.

da Silva Santos, Kleber Roberto, Emília Villani, Wesley Rodrigues de Oliveira, and Augusto Dttman. 2022. Comparison of visual servoing technologies for robotized aerospace structural assembly and inspection. *Robotics and Computer–Integrated Manufacturing* 73: 102237.

Dani, Ines, Welf-Guntram. Drossel, Nikolaus Milaev, Hannes Korn, Christian Hannemann, Joerg Hohlfeld, and Rafi Wertheim. 2020. Sustainability of industrial components using additive manufacturing and foam materials. *Procedia Manufacturing* 43: 10–17.

Darwish, Ashraf, and Aboul Ella Hassanien. 2018. Cyber physical systems design, methodology, and integration: The current status and future outlook. *Journal of Ambient Intelligence and Humanized Computing* 9: 1541–1556.

Ding, Pengfei, Xianzhen Huang, Chengying Zhao, Huizhen Liu, and Xuewei Zhang. 2023. Online monitoring model of micro-milling force incorporating tool wear prediction process. *Expert Systems with Applications* 223: 119886.

References

du Plessis, Anton, Chris Broeckhoven, Ina Yadroitsava, Igor Yadroitsev, Clive H. Hands, Ravi Kunju, and Dhruv Bhate. 2019. Beautiful and functional: A review of biomimetic design in additive manufacturing. *Additive Manufacturing* 27: 408–427.

Elzein, Basma. 2024. Nano Revolution: "Tiny tech, big impact: How nanotechnology is driving SDGs progress." *Heliyon* 10: e31393.

Essink, W. P., J. M. Flynn, S. Goguelin, and V. Dhokia. 2017. Hybrid ants: A new approach for geometry creation for additive and hybrid manufacturing. *Procedia CIRP* 60: 199–204.

Gromat, Theo, Julien Gardan, Omar Saifouni, Ali Makke, and Naman Recho. 2023. Generative design for additive manufacturing of polymeric auxetic materials produced by fused filament fabrication. *International Journal on Interactive Design and Manufacturing (IJIDeM)* 17: 2943–2955.

Groover, Mikell P. 2016. *Automation, production systems, and computer-integrated manufacturing*, 4th edn. Boston: Pearson.

Hanenkamp, Nico, and Oliver Zipse. Sustainability in manufacturing transforming. Envisioning the factory of the future. In: *Road to net zero. Strategic pathways for sustainability-driven business transformation*, ed. Oliver Zipse, Joachim Hornegger, Thomas Becker, Markus Beckmann, Michael Bengsch, Irene Feige, and Markus Schober. Springer.

ISO/ASTM. 2022. ISO/ASTM 52900:2022 (E) Fabricación aditiva. Principios generales. Fundamentos y vocabulario. ISO/ASTM International.

Ji, Wei, and Lihui Wang. 2019. Industrial robotic machining: A review. *The International Journal of Advanced Manufacturing Technology* 103: 1239–1255.

Jiang, Xiaogeng, Zhiwei Cui, Liang Wang, Chang Liu, Maojun Li, Jian Liu, and Yu Du. 2022. Critical geometric errors identification of a five-axis machine tool based on global quantitative sensitivity analysis. *The International Journal of Advanced Manufacturing Technology* 119: 3717–3727.

Khajuria, Anupam, Vella A. Atienza, Suchana Chavanich, Wilts Henning, Ishrat Islam, Ulrich Kral, Meng Liu, Xiao Liu, Indu K. Murthy, Temitope D. Timothy Oyedotun, Prabhat Verma, Guochang Xu, Xianlai Zeng, and Jinhui Li. 2022. Accelerating circular economy solutions to achieve the 2030 agenda for sustainable development goals. *Circular Economy* 1(1): 100001.

Kim, Seong Hyeon, Eunseok Nam, Tae In Ha, Soon-Hong Hwang, Jae Ho Lee, Soo-Hyun Park, and Byung-Kwon Min. 2019. Robotic machining: A review of recent progress. *International Journal of Precision Engineering and Manufacturing* 20: 1629–1642.

Kong, Xiangcheng, Jia Deng, Jingyan Dong, and Paul H. Cohen. 2020. Study of tip wear for AFM-based vibration-assisted nanomachining process. *Journal of Manufacturing Processes* 50: 47–56.

Kumar, Abhinav, Manjesh Kumar, Anupam Alok, H. N. S Yadav, and Manas Das. 2021. Fabrication of microtool for micromachining: A review. *Materials Today: Proceedings* 47: 3911–3918.

Ma, Zhongwei, Zhiwu Xu, Zhengwei Li, Shu Chen, You Wu, and Jiuchun Yan. 2024. Improving the quality of resistance welded thermoplastic composite joints by applying ultrasonic. *Composites Part B* 277: 111398.

Maamar, Zakaria, Ejub Kajan, Mohammed Al-Khafajiy, Murtada Dohan, Amjad Fayoumi, and Fadwa Yahya. 2023. A multi-type artifact framework for cyber–physical, social systems design and development. *Internet of Things* 22: 100820.

Masuzawa, T. 2000. State of the art of micromachining. *Annals CIRP* 49: 473.

McCarthy, C. T., and M. A. McCarthy. 11—Design and failure analysis of composite bolted joints for aerospace composites. In: Polymer Composites in the Aerospace Industry, ed. P. E. Irving, C. Soutis, 295–334. Woodhead Publishing. ISBN 9780857095237

Mei, Biao, Zengsheng Liang, Yuedong Xie, Yun Fu, and Yongtai Yang. 2023. Positioning accuracy enhancement of a robotic assembly system for thin-walled aerostructure assembly. *Journal of Industrial Information Integration* 35: 100518.

Mohammadi, Y., and K. Ahmadi. In-process frequency response function measurement for robotic milling. *Experimental Techniques* 47: 797–816.

Naiju, C.D. 2021. DFMA for product designers: A review. *Materials Today: Proceedings* 46: 7473–7478.

Nazir, Aamer, Ozkan Gokcekaya, Kazi M. Masum, Onur Ertugrul Billah, Jingchao Jiang, Jiayu Sun, and Sajjad Hussain. 2023. Multi-material additive manufacturing: A systematic review of design, properties, applications, challenges, and 3D printing of materials and cellular metamaterials. *Materials and Design* 226: 111661.

Nyamekye, Patricia, Pinja Nieminen, Mohammad Reza Bilesan, Eveliina Repo, Heidi Piili, Antti Salminen. 2021. Prospects for laser based powder bed fusion in the manufacturing of metal electrodes: A review. *Applied Materials Today* 23: 101040.

Pascual, Alejandro, Naiara Ortega, Soraya Plaza, Luis Norberto López de Lacalle, and Eneko Ukar. 2023. Analysis of the influence of L-PBF porosity on the mechanical behavior of AlSi10Mg by XRCT-based FEM. *Journal of Materials Research and Technology* 22: 958–981.

Pudi, Varshini, M. V. A. Raju Bahubalendurni, and Anuj Desai. 2023. Carbon foot print analysis integration with DFMA concept: A case study. *Materials Today: Proceedings* 90: 179–183.

Shekhar, Shivang, Bekir Bediz, and O. Burak Ozdoganlar. 2023. Tool-tip dynamics in micromachining with arbitrary tool geometries and the effect of spindle speed. *International Journal of Machine Tools and Manufacture* 185: 103981.

Stavropoulos, Panagiotis. 2023. *Additive manufacturing: Design, processes and applications.* Springer.

Tao, Fei, Qinglin Qi, Lihui Wang, et al. 2019. Digital twins and cyber-physical systems toward smart manufacturing and industry 4.0: Correlation and comparison. *Engineering* 5: 653–661.

Tsirkas, S.A. 2018. Numerical simulation of the laser welding process for the prediction of temperature distribution on welded aluminium aircraft components. *Optics and Laser Technology* 100: 45–56.

United Nations. 2015. *Transforming our world: The 2030 agenda for sustainable development.* Retrieved July 6, 2023, from https://www.refworld.org/docid/57b6e3e44.html.

Wang, Zhiping, Yicha Zhang, Donghua Dai, Dongdong Gu, Chaoyue Chen, Di Wang, and Alain Bernard. 2023. A direct toolpath constructive design method for controllable porous structure configuration with a TSP-based sequence planning determination. *Chinese Journal of Mechanical Engineering: Additive Manufacturing Frontiers* 2: 100063.

Wang, Lihui, Martin Törngren, and Mauro Onori. 2015. Current status and advancement of cyber-physical systems in manufacturing. *Journal of Manufacturing Systems* 37: 517–527.

Wang, Weijun, Chen Zheng, Feng Tang, and Yicha Zhang. 2021. A practical redesign method for functional additive manufacturing. *Procedia CIRP* 100: 566–570.

XiangFei, Li, Huang Tao, Zhao Huan, Zhang XiaoMing, Yan SiJie, Dai Xing, and Ding Han. 2022. A review of recent advances in machining techniques of complex surfaces. *Science China Technological Sciences* 65: 1915–1939.

Xiong, Yi, Yunlong Tang, Qi Zhou, Yongsheng Ma, and David W. Rosen. 2022. Intelligent additive manufacturing and design state of the art and future perspectives. *Additive Manufacturing* 59: 103139.

Zhang, Jiabao, Jianpeng Wang, Guoqing Zhang, Zexuan Huo, Zejia Huang, and Linjia Wu. 2024. A review of diamond synthesis, modification technology, and cutting tool application in ultra-precision machining. *Materials and Design* 237: 112577.

Zhu, Kunpeng. 2022. *Smart machining systems. Modelling, monitoring and informatics.* Springer.

Chapter 4
All About Technology

Abstract Vast advances in technology emerged during the Third Industrial Revolution. Computers, transistors, and the Internet have fostered new technological developments such as artificial intelligence (AI). AI is increasing its role in various industries, particularly in design, where its influence continues to grow. In design, technology has helped evolve traditional manual drawings to computer-aided design, engineering, and manufacturing, and, currently, more advanced digital product development approaches. The design process can be amplified using big data and artificial intelligence technologies. Nowadays, data is pivotal in industry, and design can take advantage of the vast amounts of data and proper software for processing data. As technology evolves, the role of "machines" in the design process increases, opening a debate on whether technology will substitute the role of humans in the creative process. Ethical concerns and public acceptance are essential for the development of these new technologies, though their relevance may differ across sectors. In the case of aerospace, artificial intelligence has been used in multiple applications, and it is expected that, because of the coming advances in technologies, more applications can be developed in activities such as design.

Nowadays, artificial intelligence is becoming increasingly important in many industrial activities. Artificial intelligence already significantly affects design, and its influence will continue to grow. This progress has been enabled by several key technologies that create a foundation that makes the development of different artificial intelligence applications possible.

4.1 The Foundations

In the latter half of the twentieth century, the Third Industrial Revolution introduced significant innovations that transformed economies and societies. The advent of computers, their associated devices, and the Internet played a crucial role in a profound transformation at all levels. These advances helped build the foundations for subsequent modernization.

Although humans have been inhabiting the Earth for thousands of years, their evolution has been tremendously slow and progressive. Indeed, it would be challenging to find significant differences in long periods such as the Middle Ages. Even though enormous advances were made during Ancient Greece and the Roman Empire, consolidating these changes took a long time. However, with the onset of the First Industrial Revolution, transformations have significantly accelerated. There is not only a significant difference between the economy and society of the eighteenth and nineteenth centuries, but advances that have taken place in recent decades have produced profound transformations evident in a matter of decades.

The driving force behind change is the continuous development of science and technology. Key technologies for these changes include steam machines, electricity, computers, and the Internet. At present, humanity is considered to be at the beginning of a new industrial revolution, the Fourth Industrial Revolution. At the core, this new paradigm is characterized by various technologies, such as big data and advanced data analytics, additive manufacturing, augmented and virtual reality, and the Internet of Things (Sartal et al. 2020).

The Fourth Industrial Revolution technologies are beginning to showcase the potential of artificial intelligence as creative machines, but their capabilities still need to be improved. Many of the key technologies for the development of artificial intelligence are based on advanced technologies from the Third Industrial Revolution. Discoveries made toward the middle of the twentieth century, such as the transistor, the chip, the computer, and the Internet, have been crucial in this regard (Isaacson 2014). However, new ones like quantum computing and the new 5G/6G communication networks may further enhance artificial intelligence applications.

The ability of the machines to process data is a crucial element. A conventional metric to measure the processing capability is the speed of a central processing unit (CPU), which correlates with the number of transistors on a computer chip. Technological advances increased the number of transistors included in a chip, following a trend consistent with Moore's Law (Moore 1965), predicting that the number of transistors that can be placed on a chip will double every 18–24 months (Shalley et al. 2015). The transistor count is a metric that helps understand the complexity of an integrated circuit.[1] In 1971, the Intel 4004 microprocessor had 2250 transistors in a size of 12 mm^2. In 2023, the Apple A17 had 19 billion transistors in 103.8 mm^2. Thus, the transistor count per mm^2 increased from 187.5 to 183,044,316.

---

[1] For more information: https://en.wikipedia.org/wiki/Transistor_count.

4.2 From Sketches to Digital Products

For centuries, drawings have been created using freehand techniques or drawing aids. Regarding technical or engineering drawings, until the 1970s–1980s, these drawings were made using a pencil, drawing pen, and ink, see Fig. 4.1 up_left (Mas et al. 2015; Chang 2014). This drawing set has been the standard for drawings since at least the 1500s when Leonardo da Vinci was sketching his designs, and it is also used in the early or conceptual stages of design (Dickinson et al. 2005). The drafter's knowledge was critical, as any mistake might lead to the recreation of the drawing from scratch. Moreover, spatial vision is essential as the drafter must envision the intersection of solid entities and reproduce them into an engineering drawing (Chang 2014).

The advent of computers and the advances in computer graphics (Benest 1979) have completely reshaped the world of industrial design. Computers have made it easier for designers to create models and drawings that later will be converted into prototypes, parts, and assemblies. With the aid of computers, the designers can produce drafts, modify them as many times as needed, share them to discuss with their teams, and store them to revise or reuse in the future.

Computers help define a new way to work and redefine technical drawing (Ramatsetse et al. 2023). The use of CAD software is a central part of the design process. A CAD drawing includes geometries in layers such as lines, circles, arcs, and polygons, built using essential components such as vertices, angles, radii, and so on (Lee and Kwon 2010). CAD drawings are used in many applications in aerospace, architecture, automotive, prosthetics, and shipbuilding (Lee and Kwon 2010).

CAD software has a long tradition; its roots can be traced back to the 1950s and 1960s. Patrick J. Hanratty is renowned as the father of CAD/CAM, as he conceived computer-aided design and computer-aided manufacturing, helping develop solutions such as Pronto, an early commercial numerical control programming language, and DAC (Design Automated by Computer).[2] In January 1963, Ivan Edward Sutherland at the Massachusetts Institute of Technology (MIT)[3] introduced Sketchpad, the first commercial computer-aided design (CAD) software. Sutherland worked on this innovation as part of his *Sketchpad, a Man-machine Graphical Communication System* Ph.D. thesis, presented in 1963 (Sutherland 1963). The Sketchpad (Fig. 4.1 up_right) was a graphical system that helped users draw on the screen using a light pen. In the 1960s, 3D surface geometry computation was a research topic for companies such as Citroën and Renault in France and Ford, General Motors, and Lockheed in the United States (Stark 2022). Significant were the innovations by Pierre Bézier in curves and surface representation, which resulted in the development of the UNISURF software for Renault in 1968 (Nowacki 1987). Malik et al. (2021) reviewed other notable developments from the 1960s. In 1971, Automated Drafting and Machining (ADAM), a program aimed at automating drafting and machining written in FORTRAN, was introduced. In the 1980s, AutoCAD (Autodesk) was the

---

[2] https://halloffame.tech.uci.edu/inductee/hanratty-patrick/.
[3] The Whirlwind Project at MIT notably contributed in the development of computing and graphics (Golovine 2013).

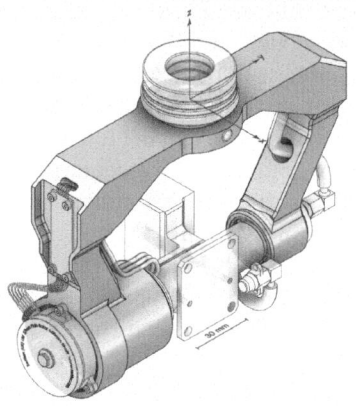

Fig. 4.1 Drawing: (up_left) Manual drawing,[4] (up_right) Ivan Sutherland using the Sketchpad,[5] (down) CAD drawing (Pilagatti et al. 2023)

first 2D CAD platform to run on desktops. 3D CAD software such as CATIA and PRO/Engineer entered the market in the same decade. Later, in the 1990s, Solid-Works was launched. Because these programs were run under Windows and were user-friendly, their use vastly increased worldwide in many sectors. An example of a CAD drawing is shown in Fig. 4.1 down.

The basic modeling techniques used in CAD are feature-based and parametric modeling (Stark 2022). Features are design entities with specific engineering meaning and relevance, which allow the user to integrate design information. In contrast, parametric modeling relies on mathematical descriptions that provide stability and are easily modified when needed. The parametric modeling approach

---

[4] By bls.gov—Drafters Occupational Outlook Handbook, 2008–09 Edition, Public Domain, https://commons.wikimedia.org/w/index.php?curid=5902438.

[5] Electronic edition of Sutherland's Sketchpad dissertation, edited by Blackwell & Rodden. Scanned by Kerry Rodden from original photograph by Ivan Sutherland, CC BY-SA 3.0. https://commons.wikimedia.org/wiki/File:SketchpadDissertation-Fig1-2.tif.

4.2 From Sketches to Digital Products

is crucial for conducting virtual development processes (Bodein et al. 2013) and generative design approaches (Jang et al. 2022).

Nowadays, the CAD software market is expected to grow at a compelling rate of over 6% from 2020 to 2030 (Lin and Lin 2023) and is dominated by a few companies such as Autodesk, Dassault Systèmes, and Siemens. In Table 4.1, it is possible to see a list of some of the most popular CAD software systems and their major characteristics. In addition, new options have been introduced in recent years, such as freeware and open-source software, such as Freecad and Onshape (Junk and Kuen 2016).

The existence of multiple CAD software programs created problems when trying to exchange programs between different platforms. The international standard ISO 10303 or STEP (STandard for the Exchange of Product model data) was introduced to address this issue (Pratt et al. 2005). Standardization is essential since various parties, within and outside the company, must use the designs in a coordinated manner. The standardization helps move toward digital product development and product lifecycle management approaches.

Computer-aided tools have increased in popularity, and several tools have been developed beyond drawing. In this sense, it is usual to refer to CAx tools (Dankwort et al. 2004). Among these, computer-aided design has been complemented by computer-aided manufacturing (CAM) and computer-aided engineering (CAE), which enable designers to create manufacturing commands for machines and simulations that improve the understanding of the behavior of designed models, respectively. Nowadays, advanced software includes all these capabilities, allowing users to transition between modules seamlessly.

The CAM tools allow the development of codes to manipulate machines, using the geometry of the CAD models as a reference. CAM systems find their roots

Table 4.1 CAD software systems

CAD software	Company	System
Creo	PTC	Parametric and direct (hybrid)
Catia	Dassault Systèmes	Parametric with a "declarative modeling" direct system (hybrid)
Fusion 360	Autodesk	Direct with a parametric component—Cloud-based (hybrid)
Inventor	Autodesk	Direct with a parametric component (hybrid)
KeyCreator	Kubotek3D	Direct 3D modeling
NX	Siemens	Parametric and direct (hybrid)
Solid Edge	Siemens	Parametric and direct (hybrid)
SolidWorks	Dassault Systèmes	Parametric feature based with direct editing capability (hybrid)
SketchUp	Trimble navigation	Direct modeler. Surface modeler. All surfaces are planar. Curved surfaces are faceted

Based on Pianca (2022)

in developments such as the first automatically controlled milling in 1952 by the Servomechanisms Laboratory at MIT. The invention used the Whirlwind computer, which allowed for translating a drawing of a machine part into a code punched out on paper tape that was later read by the milling machine. This initial development was later refined in the Automatically Programmed Tool (APT) (Benest 1979). A large number of vendors have offered CAM solutions, such as Vericut, Mastercam, MAX-PAC, NX, Catia, Delcam's PowerMill, and Open Mind's HyperMILL (Makhanov 2022; XiangFei et al. 2022).

Computer-aided engineering (CAE) is a tool for analyzing models. For instance, the models can be analyzed using finite element analysis (FEA). There are specific CAE software solutions, such as Nastran and Ansys, but there is a trend in CAD software to include CAE modules. For instance, these modules can be found in commercial software such as NX, CATIA, Pro/ENGINEER, and SolidWorks (Pan et al. 2016).

A milestone in the use of CAD software occurred in the aerospace sector. In 1988, Boeing decided to use CATIA software to design and draft the new B777 airplane, which became the first CAD-based paperless design (Jameson and Ou 2011). The project was developed under the "preferred process" (Baba and Nobeoka 1998), which included concurrent product definition, design-build teams, digital product definition, and digital pre-assembly. The definition of a digital product consists of all components using 3D CAD data. Some benefits of the project included 75% fewer design changes than in previous projects, reduced parts interference problems through digital pre-assembly, reduced engineering hours, and fewer prototypes, including mockups.

The development of the Boeing B777 using CATIA is an excellent example of the trend toward digitalizing product development. Initially, digital product development faced challenges as the number of files increased, causing designers, engineers, and analysts to spend too much time searching (Stark 2022). The Ford Motor Company identified this problem in the mid-1990s and called it "Sam Schwartz's Search for Data".

The current approach for dealing with the problem of searching for data, being more efficient, and taking advantage of the information generated through all the project stages is the product lifecycle management (PLM) approach. PLM is "a strategy for the management of all information and processes linked to the product during its life cycle, from the definition of the product requirements up to the recycling, and the context of extended enterprise" (Bodein et al. 2013). The PLM approach is an evolution from previous efforts for data management, such as engineering data management (EDM), team data management (TDM), and product data management (PDM) (Stark 2022). Product data management (PDM), as part of the PLM approach, allows a firm to create a repository for storing CAD/CAM files and revisions, specifications, documentation and standards, manufacturing information and requirements, calculations, illustrations, and suppliers details (Tasalloti et al. 2016).

Digital production development is notably driven by the shift from 2D drawing to 3D models. This movement relies on the concept of model-based definition (MBD). MBD is based on a 3D model that integrates the data needed to define a part (geometric dimensioning, tolerancing, and notes) (Castro-Cañas and Pavón-Domínguez 2023). This model serves as a reference for the model-based enterprise (MBE) approach to product development (Lubell et al. 2012; Camba et al. 2017). In terms of standardization, in 2003, the American Society of Mechanical Engineers (ASME)[6] issued the ASME Y14.41 standard that helps standardize how companies create and use their 3D models. These models can include Geometric Dimensioning and Tolerancing (GD&T) and notes (Quintana et al. 2010). This standard was transferred to the ISO 16792:2006 international standard (Frechette et al. 2013).

4.3 Artificial Intelligence

In 1995, in the book *Being Digital*, Nicholas Negroponte advanced a substantial transformation of the economy from a material-centered economy based on atoms to a new bit-centric economy (Negroponte 1995). Nowadays, the importance of information and data is evident within the context of the Fourth Industrial Revolution (Carou 2021).

In 1956, at Dartmouth College in New Hampshire, a group of computer scientists met to discuss the idea that computers could behave like humans. One of these scientists was John McCarthy, who is credited with coining the concept of artificial intelligence (Burgess 2021).

Artificial intelligence has evolved since the 1950s from classic machine learning approaches based on statistical models to the more recent deep learning strategies developed since the 1990s (Oliveira et al. 2024), being deep learning a subset of machine learning, and the latter a subset of the broader concept of artificial intelligence (Awuah et al. 2024). Many applications have extensively used machine and deep learning (Tripathi et al. 2022). Machine learning is defined "as an assortment of complex algorithms designed to emulate that of the human mind through learning from the surrounding environment and/or experiences". Meanwhile, a definition for deep learning is "a multi-layered learning module, referred to as a neural network".

There is plenty of discussion about the concept of artificial intelligence. The truth is that it relates to the idea of intelligence usually associated with humans. Even when dealing with humans, the concept of intelligence is complex. In this sense, in 1983, Gardner presented a study in which he developed the idea of multiple intelligences,[7] interpersonal, logical-mathematical, spatial, musical, bodily kinesthetic, linguistic, and intrapersonal intelligence. Later, Gardner introduced the concept of naturalistic intelligence (Weisberg 2006). In this sense, it can be reasonable to state that the intelligence associated with machines cannot be directly assimilated to human

---

[6] More updated information can be found in www.asme.org/codes-standards/y14-standards.

[7] The concept of multiple intelligences was first developed in *Frames of Mind: The Theory of Multiple Intelligences* (Gardner 1983). The concept was further developed in *Creating minds: An anatomy of creativity seen through the lives of Freud, Einstein, Picasso, Stravinsky, Eliot, Graham, and Gandhi* (Gardner 1993)

intelligence, as human intelligence should be seen as a myriad of intelligences beyond logical-mathematical intelligence (Togni et al. 2021).

In a data-dominated context, machines have a decisive advantage over humans. Machines can search for information from multiple sources practically instantaneously, and with the appropriate algorithms, they can select and process it to obtain knowledge. Certain aspects can complicate the task, such as the variety of formats, accessibility, and incomplete or mislabeled data. But today, humans cannot process as much information as machines can.

The emergence of big data present machines with enormous potential; thus, artificial intelligence finds vast opportunities to grow. Data is akin to blood for the human body. Without the necessary flow of blood, the various parts of the human body would not function properly as expected. Similarly, an artificial intelligence-based system can only realize its potential with data. In the case of artificial intelligence, the more data, the better.

Data and artificial intelligence are opening a new era for society. In 2008, Chris Anderson[8] discussed the "end of theory" in the American Wired magazine in the article *The End of Theory: the Data Deluge Makes the Scientific Method Obsolete*. Anderson took as a point for discussion the way humans create theories based on models and the way the access to data and the improvement in data analytics defies that approach: "We can throw the numbers into the biggest computing clusters the world has ever seen and let statistical algorithms find patterns where science cannot". Thus, according to Anderson, there is no need to find causation; instead, correlation can be identified simply by analyzing extensive datasets.

The supercomputers or the incipient quantum computers are great allies for artificial intelligence (Yalcin et al. 2024). The processing capacity of these computers makes it possible to tackle highly complex problems that require training large AI models (Fatima et al. 2022). Although the technology is still mainly experimental, quantum computing is expected to be a game changer for artificial intelligence. As Taylor (2020) states quantum computing and quantum artificial intelligence have the "potential to be both a disruptive innovation and a social inflection point, enabling a qualitatively different level of performance in several critical technologies, such as artificial intelligence".

The already massive use of artificial intelligence has brought several potential drawbacks of the technology to the debate. For instance, Clarke (2019) addressed five threats of artificial intelligence, specifically artifact autonomy, inappropriate assumptions about data, inappropriate assumptions about the inferencing process, the inferencing process opaqueness, and irresponsibility.

Public acceptance is a crucial consideration when it comes to using machines and artificial intelligence for creative tasks. Art is perhaps the most critical field in which it is most relevant to debate whether or not to accept the participation of artificial intelligence. Once the machines have been able to create art, the public's reaction will remain to be seen. In this sense, Chiarella et al. (2022) developed a study selecting 110 volunteers at the *ArtVerona Fiera dell'Arte*. According to the study, there is

---

[8] https://www.wired.com/2008/06/pb-theory/.

a negative bias toward machine-generated art, with the aesthetic valuation tending to be lower than that given to works of art created by humans. In other words, there are prejudices against the art created by machines. A similar negative bias toward computer-generated music has been identified, especially when musicians participate in the study (Moffat and Kelly 2006). However, people's opinions tend to be more positive regarding collaborative efforts between humans and machines (Messer 2024).

Artificial intelligence must find patterns in the data to replicate or generate something. With this philosophy in mind, artificial intelligence operates based on existing information. Humans work similarly based on what already exists, but they can also depart from what already exists or adapt concepts from unrelated fields of knowledge. Thus, there is a significant difference between human and artificial intelligence. The latter is based on programmed algorithms. Algorithms are nothing more than rules that can be highly complex. Although humans also follow the rules, transformative and disruptive creativity emerges when established rules are bent, altered, or exceeded. That is when truly new ideas emerge.

There is an ongoing discussion about whether machines can be creative or not. Likely, the problem itself is trying to compare machines and humans. Colton and Wiggins noted, "Computational systems are not human, and so the creativity they exhibit will be creativity, but not as we know it: never exactly the same as in humans". This starting point allows them to refer to a specific kind of creativity, computational creativity that can be defined as "the philosophy, science and engineering of computational systems which, by taking on particular responsibilities, exhibit behaviors that unbiased observers would deem to be creative". Ultimately, there is an ongoing debate about whether machines are superior to humans in certain areas. In this sense, Makoto Sei Watanabe stated that "machines are better than people at solving complex problems with many intertwined conditions. In that realm, people are no match for machines. But people are the only ones who can create an image that does not yet exist. Machines do not have dreams" (Leach 2022).

Although artificial intelligence is still an emerging technology, it has already made significant strides in the aerospace industry. It offers promising applications such as reducing environmental impact, fostering smart manufacturing, and enhancing production and product support phases (Le Clainche et al. 2023). These advancements inspire optimism about the future of artificial intelligence in this field.

References

Awuah, Wireko Andrew, Favour Tope Adebusoye, Jack Wellington, Lian David, Abdus Salam, Amanda Leong Weng Yee, Edouard Lansiaux, Rohan Yarlagadda, Tulika Garg, Toufik Abdul-Rahman, Jacob Kalmanovich, Goshen David Miteu, Mrinmoy Kundu, and Nikitina Iryna Mykolaivna. 2024. Recent outcomes and challenges of artificial intelligence, machine learning, and deep learning in neurosurgery. *World Neurosurgery: X* 23: 100301.

Baba, Yasunori, and Kentaro Nobeoka. 1998. Towards knowledge-based product development: The 3-D CAD. *Research Policy* 26: 643–659.

Benest, I.D. 1979. A review of computer graphics publications. *Computers and Graphics* 4: 95–136.

Bodein, Yannick, Bertrand Rose, and Emmanuel Caillaud. 2013. A roadmap for parametric CAD efficiency in the automotive industry. *Computer-Aided Design* 45: 1198–1214.

Burgess, Matt. 2021. Artificial Intelligence. How machine learning will shape the next decade. *Wired*.

Camba, Jorge D., Manuel Contero, Pedro Company, and David Pérez. 2017. On the integration of model-based feature information in product lifecycle management systems. *International Journal of Information Management* 37: 611–621.

Carou, Diego. 2021. *Aerospace and digitalization. A transformation through key industry 4.0 technologies*, ed. J. P. Davim. Springer Nature.

Castro-Cañas, Luis, and Pablo Pavón-Domínguez. 2023. Dimensioning Method for 3D Modelling. In: *Advances in Design Engineering III. Proceedings of the XXXI INGEGRAF International Conference*, 29–30 June, 1 July 2022, Málaga, Spain, ed. Francisco Cavas-Martínez, Manuel D. Marín Granados, Ramón Mirálbes Buil, and Oscar D. de-Cózar-Macías. Springer.

Chang, Kuang-Hua. 2014. *Product design modeling using CAD/CAE*. The Computer Aided Engineering Design Series. Springer.

Chiarella, Salvatore G., Giulia Torromino, Dionigi M. Gagliardi, Dario Rossi, Fabio Babiloni, and Giulia Cartocci. 2022. Investigating the negative bias towards artificial intelligence: Effects of prior assignment of AI-authorship on the aesthetic appreciation of abstract paintings. *Computers in Human Behavior* 137: 107406.

Clarke, Roger. 2019. Why the world wants controls over artificial intelligence. *Computer Law and Security Review* 35: 423–433.

Colton, Simon, and Geraint A. Wiggins. Computational creativity: The final frontier?

Dankwort, C. Werner, Roland Weidlich, Birgit Guenther, and Joerg E. Blaurock. 2004. Engineers' CAx education—it's not only CAD. *Computer-Aided Design* 36: 1439–1450.

De Togni, Giulia, Sonja Erikainen, Sarah Chan, and Sarah Cunningham-Burley. 2021. What makes AI 'intelligent' and 'caring'? Exploring affect and relationality across three sites of intelligence and care. *Social Science and Medicine* 277 (2021): 113874.

Dickinson, John K., Zhisong Yu, Yong Zeng, and Helder Antunes. 2005. Pen–tablet as a CAD interface alternative. *Robotics and Computer-Integrated Manufacturing* 21: 465–474.

Fatima, Samar, Kevin C. Desouza, Gregory S. Dawson, and James S. Denford. 2022. Interpreting national artificial intelligence plans: A screening approach for aspirations and reality. *Economic Analysis and Policy* 75: 378–388.

Frechette, S. P., A. T. Jones, B. R. Fischer. 2013. Strategy for testing conformance to geometric dimensioning & tolerancing standards. *Procedia CIRP* 10: 211–215. https://doi.org/10.1016/j.procir.2013.08.033

Golovine, Jean Claude. 2013. Experimental user interface design toolkit for interaction research (IDTR). Ph.D. Thesis. Robert Gordon University.

Isacsson, Walter. 2014. *The innovators: How a group of inventors, hackers, geniuses and geeks created the digital revolution*. Simon and Schuster.

Jameson, Antony, and Kui Ou. 2011. 50 years of transonic aircraft design. *Progress in Aerospace Sciences* 47: 308–318.

Jang, Seowoo, Soyoung Yoo, and Namwoo Kang. 2022. Generative design by reinforcement learning: Enhancing the diversity of topology optimization designs. *Computer-Aided Design* 146: 103225.

Junk, Stefan, and Christian Kuen. 2016. Review of open source and freeware CAD systems for use with 3D-printing. *Procedia CIRP* 50: 430–435.

Le Clainche, Soledad, Esteban Ferrera, Sam Gibson, Elisabeth Cross, Alessandro Parente, and Ricardo Vinuesa. 2023. Improving aircraft performance using machine learning: A review. *Aerospace Science and Technology* 138: 108354.

Leach, Neil. 2022. In the mirror of AI: What is creativity? *Architectural Intelligence* 1: 15.

Lee, Suk-Hwan, and Ki-Ryong Kwon. 2010. CAD drawing watermarking scheme. *Digital Signal Processing* 20: 1379–1399.

References

Lin, Yi-Hsin, and Feng-Jyh Lin. 2023. An assessment framework for the purchase of 3D CAD software in manufacturing industries. *Technological Forecasting and Social Change* 192: 122573.

Lubell, Joshua, Kenway Chen, John Horst, Simon Frechette, and Paul Huang. 2012. Model based enterprise/technical data package summit report (NIST technical note 1753). NIST Technical Note, 1753.

Makhanov, Stanislav S. 2022. Vector fields for five-axis machining. A survey. *The International Journal of Advanced Manufacturing Technology* 122: 533–575.

Malik, Shubhit, Vineet Kumar, and Pradeep Gahlot. 2021. An analogy between conventional and unorthodox CAD methods. *Materials Today: Proceedings* 43: 2098–2104.

Mas, F., R. Arista, M. Oliva, B. Hiebert, I. Gilkerson, and J. Rios. 2015. A review of PLM impact on US and EU Aerospace Industry. *Procedia Engineering* 132: 1053–1060.

Messer, Uwe. 2024. Co-creating art with generative artificial intelligence: Implications for artworks and artists. *Computers in Human Behavior: Artificial Humans* 2 (1): 100056.

Moffat, D.C., and M. Kelly. 2006. An investigation into people's bias against computational creativity in music composition. *Assessment* 13 (11): 1–8.

Moore, G.E. 1965. Cramming more components onto integrated circuits. *Electronics* 38(8).

Negroponte, Nicholas. 1995. *Being digital*. New York: Alfred A, Knopf.

Nowacki, Horst. 1987. Honoring Pierre Bézier. *Computer Aided Geometric Design* 4 (3): 167–170.

Oliveira, Franklin, Daniel G. Costa, Flávio. Assis, and Ivanovitch Silva. 2024. Internet of Intelligent Things: A convergence of embedded systems, edge computing and machine learning. *Internet of Things* 26: 101153.

Pan, Zhiyi, Xin Wang, Rumin Teng, and Xuyang Cao. 2016. Computer-aided design-while-engineering technology in top-down modeling of mechanical product. *Computers in Industry* 75: 151–161.

Pianca, Eddi. 2022. The embedded design process: CAD/CAM and prototyping. In: *Foundations of robotics. A multidisciplinary approach with Python and ROS*, ed. Damith Herath and David St-Onge. Springer.

Pratt, Michael J., Bill D. Anderson, and Tony Ranger. 2005. Towards the standardized exchange of parameterized feature-based CAD models. *Computer-Aided Design* 37: 1251–1265.

Pilagatti, Adriano Nicola, Eleonora Atzeni, and Alessandro Salmi. 2023. Exploiting the generative design potential to select the best conceptual design of an aerospace component to be produced by additive manufacturing. *The International Journal of Advanced Manufacturing Technology* 126: 5597–5612.

Quintana, Virgilio, Louis Rivest, Robert Pellerin, Frédérick. Venne, and Fawzi Kheddouci. 2010. Will Model-based Definition replace engineering drawings throughout the product lifecycle? A global perspective from aerospace industry. *Computers in Industry* 61: 497–508.

Ramatsetse, Boitumelo, Ilesanmi Daniyan, Khumbulani Mpofu, and Olasumbo Makinde. 2023. State of the art applications of engineering graphics and design to enhance innovative product design: A systematic review. *Procedia CIRP* 119: 699–709.

Sartal, Antonio, Diego Carou, and J. Paulo Davim. eds. 2020. *Enabling technologies for the successful deployment of Industry 4.0*, 1st edn. Boca Ratón: CRC Press.

Shalley, Christina E., Michael A. Hitt, and Jing Zhou. 2015. *The Oxford handbook of creativity, innovation, and entrepeneurship*. Oxford University Press.

Stark, Rainier. 2022. *Virtual product creation in industry. The difficult transformation from it enabler technology to core engineering competence*. Springer.

Sutherland, Ivan Edward. 1963. Sketchpad, a man-machine graphical communication system. Massachusetts Institute of Technology Library.

Tasalloti, H., H. Eskelinen, P. Kah, and J. Martikainen. 2016. An integrated DFMA–PDM model for the design and analysis of challenging similar and dissimilar welds. *Materials and Design* 89: 421–431.

Taylor, Richard D. 2020. Quantum artificial intelligence: A "precautionary" U.S. approach? *Telecommunications Policy* 44: 101909.

Tripathi, Satvik, Alisha Isabelle Augustin, Adam Dunlop, Rithvik Sukumaran, Suhani Dheer, Alex Zavalny, Owen Haslam, Thomas Austin, Jacob Donchez, Pushpendra Kumar Tripathi, and Edward Kim. 2022. Recent advances and application of generative adversarial networks in drug discovery, development, and targeting. *Artificial Intelligence in the Life Sciences* 2: 100045.

Weisberg, Robert W. 2006. *Creativity understanding innovation in problem solving, science, invention, and the arts*. Wiley.

XiangFei, Li, Huang Tao, Zhao Huan, Zhang XiaoMing, Yan SiJie, Dai Xing, and Ding Han. 2022. A review of recent advances in machining techniques of complex surfaces. *Science China Technological Sciences* 65: 1915–1939.

Yalcin, Haydar, Tugrul Daim, Mahdieh Mokhtari Moughari, and Alain Mermoud. 2024. Supercomputers and quantum computing on the axis of cyber security. *Technology in Society* 77: 102556.

Chapter 5
Structural Optimization

Abstract Traditional design was mainly focused on functionality. Thus, designers had to define the geometry to meet objectives such as strength or thermal needs. Structural optimization techniques were developed to create efficient structures. The emergence of computational solutions has helped evolve structural optimization, with size, shape, and topology optimization already well-established methods that serve different needs, such as modifying dimensions, contours, and creating more complex geometries by optimizing material distribution within a design space, respectively. Topology optimization is an advanced method that helps generate innovative and lightweight structures. It is, therefore, a method more closely related to generative design. The method has been proven effective in various industries, including aerospace.

The product's functional requirements mainly drove traditional design methods. Thus, the geometry was defined to meet the objective, satisfying, among other requirements, strength and thermal requirements. However, optimizing the geometry of parts and products is also required in modern manufacturing. This approach aims to develop efficient structures.

Structural optimization has gained significant relevance in recent decades, though Karl Culmann had already introduced the idea of optimizing the mass of structures in civil engineering in 1866 through his studies on structural analysis (Culmann 1866; Stoiber and Kromoser 2021). One of the earliest works that laid the foundation for structural optimization dates back to the early twentieth century. In this sense, Michell's truss theory helps to find a frame with the optimal structure (Michell 1904; Luo et al. 2024). However, these early theoretical approaches lacked the computational tools to solve complex geometries.

It was decades after the work by Michell when advanced numerical and computational methods (Bendsøe and Sigmund 2004) renewed interest in structural optimization. In 1988, Bendsøe and Kikuchi presented the study *Generating Optimal Topologies in Structural Design Using a Homogenization method* describing the homogenization method (Bendsøe and Kikuchi 1988). This study helps search for an "optimal distribution in space of an anisotropic material that is constructed by introducing an infimum of periodically distributed small holes in a given homogeneous, isotropic material, with the requirement that the resulting structure can carry the given loads as well as satisfy other design requirements".[1]

Structural optimization aims at "finding an optimal configuration within a given design domain for specific objectives, subject to constraints, loads, and boundary conditions" (de Andrade and Magalhães 2022). This optimization process is a mathematical problem that involves minimizing the $f(x, y)$ objective function, as described by Christensen and Klarbring (2009). The problem has the following elements in its general form:

- Objective function (f): a function used to classify designs (i.e., their goodness). Usually, this function is designed to minimize a problem such as weight, displacement in a given direction, effective stress, or even the cost of production.
- Design variable (x): a function or vector that describes the design, for instance, geometry or material.
- State variable (y): a function or vector representing the structure's response, including parameters such as displacement, stress, strain, or force.

5.1 Size, Shape, and Topology Optimization

The first approaches to performing structural optimization were size, shape, and topology optimization (Birosz et al. 2023). These methods differ; thus, it is essential to reflect on the underlying principles of each to distinguish between them effectively. The following discussion is based on the works published by Mao et al. (2023) and Martorelli and Gloria (2023).

In size optimization, the dimensions of the structure are the design variables. The optimization process then focuses on the cross-sectional area, aiming to create optimized solutions regarding mass, experienced stresses, strain distributions, and other relevant factors. In shape optimization, the design variables are the coordinates of the structure's feature points or contour features. The position of the nodes is defined to satisfy the given criteria and design objectives (e.g., reduction of stress concentration and stress shielding effects, improvement in fatigue life). Therefore, the resulting geometry generated using size and shape optimization can differ substantially.

Size and shape optimization are not adequate methods when there is a need to introduce or remove geometry to reach an optimal solution. In such cases, topology optimization is an advanced methodology that further enhances optimization. The methodology uses the layout of the local nodes as design variables, the connection relationships between nodes, and the number and position of holes. Thus, topology optimization allows for obtaining more complex geometries that can deviate significantly from the original geometry.

---

[1] To put this work in context, this study predates the introduction of the Intel Pentium Pro, Microsoft Windows 95, and Apple's Macintosh by several years.

5.1 Size, Shape, and Topology Optimization

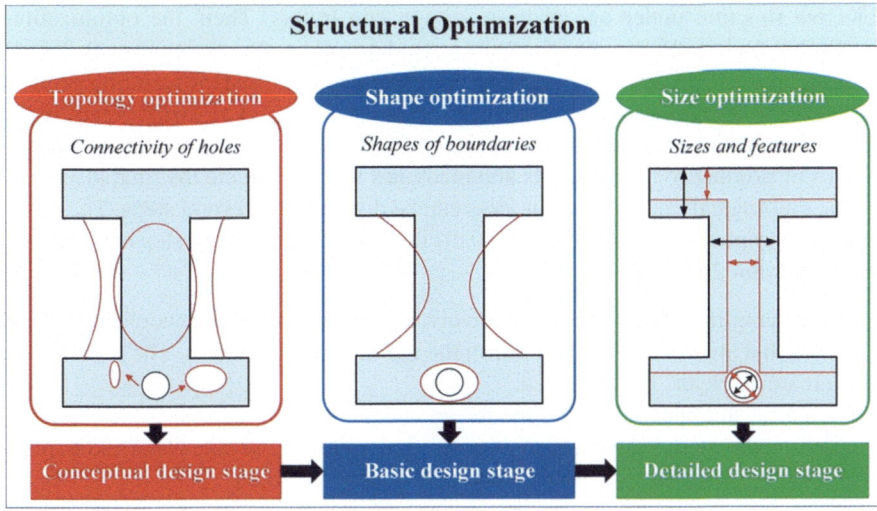

Fig. 5.1 Optimization techniques: topology optimization, shape optimization, size optimization (Gao et al. 2020)

In summary, Fig. 5.1 visually compares the three methods. Size optimization maintains the shape of the former structure but changes the size, while shape optimization allows for changing the shape's boundary. Topology optimization goes further, enabling a better utilization of the design space to obtain the optimal distribution with less attachment to the original geometry.

The aforementioned methods are distinct approaches that can be used to pursue various objectives. Thus, designers can apply these three methods differently depending on their prior knowledge of the part or product they are developing and the specific requirements they aim to meet.

According to Gao et al. (2020), topology optimization is best suited as a methodology to create starting concepts. Then, shape and size optimization can be used for basic and detailed design (see Fig. 5.1). Thus, the less the designer knows about the part or product to manufacture, the more suitable topology optimization will be to explore potential outcomes. Later, these outcomes can be fine-tuned using shape and/or size optimization.

Topology optimization is a significant methodology within this book's context as it is suitable for taking advantage of contemporary simulation tools enhanced by data analytics and advanced manufacturing processes. Moreover, it is aligned with sustainable production objectives such as mass reduction. In particular, topology optimization is a method more connected to a divergent creative process.

Marjan and Huang (2023) proposed the following definition of topology optimization: "a computational method used to generate an optimal distribution of material to achieve lightweight structures without compromising on structural integrity". Topology optimization uses finite element analysis as a basis. The material is distributed in the design domain and divided into finite elements to minimize the objective function under one or more design constraints. Then, the optimization process helps classify the finite elements as void or solid material (Jeong et al. 2023).

One of the critical drawbacks of topology optimization is that the resulting geometries must be later redesigned for manufacturability (Essink et al. 2017). Thus, these structures must be modified later to include all manufacturing constraints and adapt the part or product to the materials and machines that will create the final result.

The topological optimization process can be divided into several steps. The usual workflow proposed by Diegel et al. (2020) includes the following steps (Srivastava and Kawakami 2023):

- Simplifying the model. This step involves removing all elements of the former design that are not needed to perform the topology optimization. The objective is to create a "clean" design space.
- Assigning suitable material to the model.
- Dividing the model. A space in which the optimization can take place and a space in which the geometry cannot be altered must be identified.
- Establishing different scenario setups. The load case and constraints that the model needs to satisfy must be defined.
- Running the process. The software performs topological optimization based on mathematical algorithms to help obtain a solution that meets the specified requirements.
- Post-processing. After obtaining a solution, the model can be redefined, considering factors such as manufacturability.

5.2 Optimization Methods

Bendsøe and Kikuchi's (1988) work on the homogenization method paved the way for further developments in material distribution and structural optimization. Later, Bendsøe and Sigmund (1999) presented a density-based approach in which the density distribution problem is parameterized using a continuous density. The objective was to obtain a black-and-white structure (material and voids) using a penalization method (Luo et al. 2024). Other methods were also developed using the boundaries to describe the geometry (van Dijk et al. 2013).

Since the 1980s, several topology optimization methods have been proposed. By way of introduction, Li et al. (2022) listed some of the most used optimization methods, including the density-based method (including the solid isotropic material with penalization (SIMP) method and the rational approximation of material properties (RAMP) method), the level-set method, the topological derivative method, the evolutionary method (including evolutionary structural optimization (ESO) method and the bi-directional ESO method), the phase field method, and the emerging field of deep learning–based topology optimization method.

When comparing methods such as the solid isotropic material with penalization, the level-set method, bi-directional evolutionary structural optimization, and the ground structure method, it is interesting to note that although the methods differ

5.3 Topology Optimization Examples

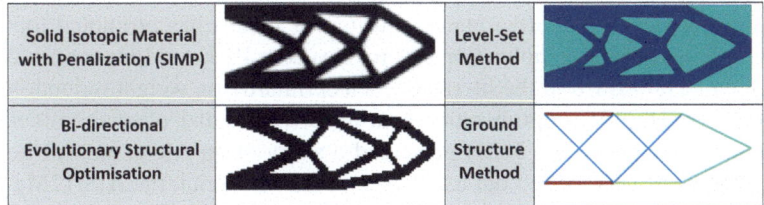

Fig. 5.2 Examples of optimization methods: (up_left) solid isotopic material with penalization (SIMP), (up_right) level-set method, (down_left) bi-directional evolutionary structural optimization; and, (down_right) ground structure method (Watson et al. 2022)

from each other, they achieve similar topologically optimal solutions, as shown in Fig. 5.2 (Watson et al. 2022).

Topology optimization traditionally relies on finite element analysis using iterative processes (Jeong et al. 2023), and it has a substantial computational cost. In recent years, some efforts have been made to use artificial intelligence techniques, particularly deep learning, in topological optimization problems. Deep learning is recognized as a tool capable of improving efficiency, yet its widespread applicability in real-world engineering remains to be fully validated (Wang et al. 2024). Various approaches that are currently under evaluation, such as the use of the convolutional neural network (CNN), convolutional neural network + conditional generative adversarial network (CNN + cGAN), ResUnet, deep belief network, GAN, physics-informed neural network (PINN), deep neural network, CNN (Unet), graph neural network and diffusion neural network (Wang et al. 2024). Applying deep learning can provide optimization methods with advantages, such as accelerating the initial optimization steps (Kallioras and Lagaros 2020) or getting optimized structures without any iteration (Yu et al. 2019).

5.3 Topology Optimization Examples

The introduction of software for topology optimization has driven the application of the methodology in different sectors. It is worth noting that this software was available in the late 1980s and early 1990s. For instance, Quint Co. (Japan) released Optishape, Altair Computing (USA) developed Optistruct, and MSC Software (USA) and CES Eckard GmbH (Germany) developed Construct and Catopo, respectively (Bendsøe and Sigmund 2004).

Topology optimization has been extensively applied in aerospace, and a few examples are presented for illustrative purposes. Some applications include layout designs for airframe structures, stiffener ribs for aircraft panels, multi-component layout designs for aerospace structural systems, and multi-fastener designs for assembled aircraft structures (Chinchanikar and Shaikh 2022). Another example is the A380 droop nose ribs and cockpit windshield design studies (Meng et al. 2020).

The development of additive manufacturing for metals has promoted topology optimization in aerospace applications, as reviewed by Blakey-Milner et al. (2021). Different cases published in the literature focus primarily on weight reduction. Yan et al. (2023) used topology optimization to design a turbine disk of an aircraft engine. To conclude, Hanush et al. (2022) presented a study of an Alcoa aircraft bearing bracket. The work was carried out using Ansys. As a material, the AlSi12Mg alloy was selected due to its excellent corrosion resistance and lightweight properties. The CAD design was modeled in Solidworks 2017, and the FEA analysis was performed using Ansys 2020 R2 Additive. Meshing was done using tetrahedral elements with an element size of 2.9 mm. Regarding the algorithms, the authors highlighted that the level-set-based topology optimization algorithms performed better compared to the density-based approach. The topology-optimized solution allowed for a 44% reduction in weight with a safety factor between 2 and 2.3 for all loading conditions.

References

Bendsøe, Martin Philip, and Noboru Kikuchi. 1988. Generating optimal topologies in structural design using a homogenization method. *Computer Methods in Applied Mechanics and Engineering* 71(2): 197–224.

Bendsøe, M. P., and O. Sigmund. 2004. *Topology optimization. Theory, methods, and applications.* Springer.

Bendsøe, M.P., and O. Sigmund. 1999. Material interpolation schemes in topology optimization. *Archive of Applied Mechanics* 69: 635–654.

Birosz, Marton Tamas, Janos Gyorgy Batorfi, and Matyas Ando. 2023. Extending the usability of the force-flow based topology optimization to the process of generative design. *Meccanica* 58: 607–618.

Blakey-Milner, Byron, Paul Gradl, Glen Snedden, Michael Brooks, Jean Pitot, Elena Lopez, Martin Leary, Filippo Berto, and Anton du Plessis. 2021. Metal additive manufacturing in aerospace: A review. *Materials and Design* 209: 110008.

Chinchanikar, Satish, and Avez A. Shaikh. 2022. A review on machine learning, big data analytics, and design for additive manufacturing for aerospace applications. *JMEPEG* 31: 6112–6130.

Christensen, Peter W., and Anders Klarbring. 2009. *An introduction to structural optimization.* Springer.

Culmann, C. 1866. Die graphische Statik. Zürich: Meyer & Zeller. https://reader.digitale-sammlungen.de/de/fs1/object/display/bsb10080652_00575.html.

de Andrade, Fábio Morais, and Cristina Almeida Magalhães. 2022. Evolutionary structural optimization in energy absorption structures. *Advances in Engineering Software* 169: 103145.

Diegel, Olaf, Axel Nordin, and Damien Motte. 2020. *A practical guide to design for additive manufacturing*, vol. 68. Singapore: Springer.

Essink, W. P., J. M. Flynn, S. Goguelin, and V. Dhokia. 2017. Hybrid ants: A new approach for geometry creation for additive and hybrid manufacturing. *Procedia CIRP* 60: 199–204.

Gao, Jie, Mi Xiao, Yan Zhang, and Liang Gao. 2020. A comprehensive review of isogeometric topology optimization: Methods, applications and prospects. *Chinese Journal of Mechanical Engineering* 33 (87).

Hanush, S. Satya, M. Manjaiah. 2022. Topology optimization of aerospace part to enhance the performance by additive manufacturing process. *Materials Today: Proceedings* 62: 7373–7378. https://doi.org/10.1016/j.matpr.2022.02.074

References

Jeong, Hyogu, Chanaka Batuwatta-Gamage, Jinshuai Bai, Yi Min Xie, Charith Rathnayaka, Ying Zhou, and YuanTong Gu. 2023. A complete physics-informed neural network-based framework for structural topology optimization. *Computer Methods in Applied Mechanics and Engineering* 417: 116401.

Kallioras, N.A., and N.D. Lagaros. 2020. DzAIN: Deep learning based generative design. *Procedia Manufacturing* 44: 591–598.

Li, Hao, Tsuguo Kondoh, Pierre Jolivet, Kozo Furuta, Takayuki Yamada, Benliang Zhu, Kazuhiro Izui, and Shinji Nishiwaki. 2022. Three-dimensional topology optimization of a fluid–structure system using body-fitted mesh adaption based on the level-set method. *Applied Mathematical Modelling* 101: 276–308.

Luo, Kaiming, Xuhui He, and Haiquan Jing. 2024. Topology optimization of bridges under random traffic loading using stochastic reduced-order models. *Probabilistic Engineering Mechanics* 75: 103583.

Mao, Bochun, Yajing Tian, Chengxin Wang, Dawei Liu, Yanheng Zhou, and Jing Li. 2023. The application of optimization design in stomatology: A literature review. *Medicine in Novel Technology and Devices* 19: 100252.

Marjan, Ali, and Luofeng Huang. 2023. Topology optimisation of offshore wind turbine jacket foundation for fatigue life and mass reduction. *Ocean Engineering* 289: 116228.

Martorelli, Massimo, and Antonio Gloria. 2023. Strategies and generative design towards the development of innovative products. In: *Springer handbook of additive manufacturing*, ed. Eujin Pei, Alain Bernard, Dongdong Gu, Christoph Klahn, Mario Monzón, Maren Petersen, and Tao Sun. Springer, 2023.

Meng, Liang, Weihong Zhang, Dongliang Quan, Guanghui Shi, Lei Tang, Yuliang Hou, Piotr Breitkopf, Jihong Zhu, and Tong Gao. 2020. From topology optimization design to additive manufacturing: Today's success and tomorrow's roadmap. *Archives of Computational Methods in Engineering* 27: 805–830.

Michell, A.G.M. 1904. The limits of economy of material in frame structures. *Philosophical Magazine* 8: 589–597.

Srivastava, Jagriti, and Hiroshi Kawakami. 2023. Systematic Review of Difference Between Topology Optimization and Generative Design. *IFAC-PapersOnLine* 56 (2): 6561–6568.

Stoiber, Nadine, and Benjamin Kromoser. 2021. Topology optimization in concrete construction: A systematic review on numerical and experimental investigations. *Structural and Multidisciplinary Optimization* 64: 1725–1749.

van Dijk, N.P., K. Maute, M. Langelaar, and F. van Keulen. 2013. Level-set methods for structural topology optimization: A review. *Struct Multidisc Optim* 48: 437–472.

Wang, Dalei, Yun Ning, Cheng Xiang, and Airong Chen. 2024. A two-stage network framework for topology optimization incorporating deep learning and physical information. *Engineering Applications of Artificial Intelligence* 133: 108185.

Watson, Marcus, Martin Leary, and Milan Brandt. 2022. Generative design of truss systems by the integration of topology and shape optimization. *The International Journal of Advanced Manufacturing Technology* 118: 1165–1182.

Yan, Cheng, Ce Liu, Han Du, Cunfu Wang, and Zeyong Yin. 2023. Topology optimization of turbine disk considering maximum stress prediction and constraints. *Chinese Journal of Aeronautics* 36 (8): 182–206.

Yu, Y., T. Hur, J. Jung, and I.G. Jang. 2019. Deep learning for determining a near-optimal topological design without any iteration. *Structural and Multidisciplinary Optimization* 59 (3): 787–799.

Chapter 6
Generative Design and Its Applications

Abstract Generative design is a powerful tool that offers designers and engineers a new way to explore the design space through a collaborative human–machine process. An analogy can be drawn with the underlying philosophy of the Flatwriter machine proposed in 1970 by architect Yona Friedman to help users generate housing layouts and furniture arrangements. Today, generative design uses load cases, materials, constraints, manufacturing methods, and objectives to create multiple solutions rather than a single fixed outcome. The generated solutions are often unlikely to be designed by human designers and resemble biological structures. Generative design is gaining attention in various sectors, such as aerospace. This chapter presents several examples of generative design including aircraft bearing brackets, drone frames, and wing ribs, and examines the foundations of this methodology.

The Hungarian-French architect Yona Friedman presented a design for a machine named Flatwriter at the Osaka World's Fair in 1970, later featured in the book *Pour une Architecture Scientifique* (1971) (Vardouli 2023; Sánchez-Pérez et al. 2022). The machine, conceived with 53 keys, allowed users to generate multiple combinations of room shapes and furniture arrangements. According to Friedman, the machine could help users choose from three to five million flats. Depending on the selection process, Friedman considered the Flatwriter a tool to guide users, giving users instructions or warnings.[1]

Although the Flatwriter was conceived for the Osaka fair, the machine was never displayed at the event. Still, the idea offers an excellent example of how human–machine cooperation was conceived more than fifty years ago, as also discussed by the visionary Nicholas Negroponte in his work *The Architecture Machine*. Later, Negroponte and Friedman cooperated in developing the YONA software for "assisted design for non-expert users" at the Media Lab, Massachusetts Institute of Technology

---

[1] Friedman explained the concept of the Flatwriter in an interview in the Office de radiodiffusion-télévision française.
https://fresques.ina.fr/europe-des-cultures-en/fiche-media/Europe00061/yona-friedman-regarding-the-machine-that-invents-flats.html.

(Sánchez-Pérez et al. 2022). This work incorporated some of the concepts developed by Friedman for the Flatwriter.

Friedman's ideas on cooperation between humans and machines connect well with current trends, such as the need for mass customization, which markets require in today's economy. Users wanted to differentiate their products from those of others, and the Flatwriter allowed for that. Moreover, it is interesting that the process is imagined as guided, not unidirectional. Thus, users receive feedback and can then modify their choices (Sánchez-Pérez et al. 2021).

Fifty years later, advances in computer science and generative design have enabled improved human-machine interaction. In this new way to cooperate, humans must provide some essential inputs to the computer, which, after processing, will provide users with some results or warnings when the input choices do not allow for obtaining a solution (i.e., there is no convergence). This new philosophy will enable computers to explore the design space and help users get more complex and efficient results. In addition, this approach shortens the process while considering numerous design factors.

6.1 The Concept

Generative design is a new methodology capable of producing results that are hardly imaginable by a human designer. It begins with a set of geometries and constraints, and, using algorithms, the program generates a range of viable solutions that meet all requirements. According to Pilagatti et al. (2023), generative design has no precise and consistent definition. However, some authors have tried defining the concept, as demonstrated in the following examples:

- "Generative design systems are aimed at creating new design processes that produce spatially novel yet efficient and buildable designs through exploitation of current computing and manufacturing capabilities" (Shea et al. 2005).
- "Generative design is the methodology for automatic creation of a large number of designs via an iterative algorithmic framework while respecting user-defined criteria and limitations" (Kallioras and Lagaros 2020).
- "Generative design represents an algorithmic approach to the classic design process. Rather than rely on the designer's creative instincts, the solution form is generated computationally by performance-driven algorithms" (Watson et al. 2023).
- "Generative design is a modern method of comprehensive design. It achieves the goal by exploring the entire design space, for example, to achieve the minimum structural compliance given a set of functional requirements and constraints based on geometry, material specifications and manufacturing considerations" (Wang et al. 2021a).

The previous definitions recognize generative design as a novel concept. As Wang et al. (2021a) stated, it is "a modern method of comprehensive design". Thus, it is not merely novel; it is a comprehensive method. In this sense, generative design is a complex concept that challenges traditional design practices. The methodology can use many variables as inputs (i.e., geometry, material specifications, and manufacturing) to explore the design space and search for solutions. Remarkably, the generative design allows for the inclusion of design constraints and performance requirements directly within the generative algorithm. Thus, the manual iteration process of the traditional design is notably reduced (HBR 2018).

Oyesola et al. (2019) characterized the design methods into three broad categories depending on the objective pursued in the process. The authors identified design for the generation of ideas for concept design (brainstorming, fishbone diagram), design for concept design evaluation (Pugh's matrix, morphological chart, checklist, etc.), and design for a specific design purpose (design for assembly, design for manufacture, quality function deployment, etc.).

The previous methods let one consider where to place generative design within these design methodologies. In this sense, generative design is a tool for developing concepts (i.e., generating ideas). However, the methodology also evaluates the proposed solution, for instance, when defining the safety factor or establishing the load case. The design will be reliable in guaranteeing the required safety factor and supporting the load case. In addition, generative design can be defined to meet a specific design criterion. For instance, it is possible to establish a manufacturing process, such as additive manufacturing or a particular type of machining. Thus, the generative process performs a design-for-manufacture strategy. Moreover, costs can also be analyzed using generative design. Based on this, it is posited that generative design extends far beyond a simple methodology for generating concepts.

Generative design is a tool that even challenges the shared idea of conceptual design. By way of illustration, Vuletic et al. (2018) stated the following regarding conceptual design: "the design is changing frequently and evolving, and the focus on detail is not as necessary as the need to generate and manipulate ideas quickly. Designers create initial solutions, then modify or combine them and create concept variants to match the design requirements, placed on the product by either customers or the context of the product being designed". In contrast, generative design often yields final or near-final manufacturing-ready results that are fully optimized. Moreover, these outcomes are especially suited to be manufactured with additive manufacturing, as this technology allows the manufacture of the parts almost without redesign. Vuletic et al. (2018) underscored the necessity of "concrete, precise and quantitative design information as an input that is often not available at the conceptual design stage" for effective generative design. However, this requirement is typically needed in generative design, in which designers must have clear insights into materials, load cases, constraints, preserve and obstacle geometries, etc., to take full advantage of this design approach.

Maurice Conti is an innovation executive who helped companies such as Airbus, Autodesk, Disney, Google, Nike, and Tesla develop new products and services.[2] In

---

[2] www.mauriceconti.com.

April 2016, Conti delivered a TED Talk titled *The Incredible Inventions of Intuitive AI*, in which he analyzed the leap from passive design to generative design. In his opinion, the impact of this shift in the next 20 years will surpass that of the previous 2000 years. A similar opinion was shared by Jeff Kowalski, Chief Technology Officer (CTO) of Autodesk, in the promotional video *The Future of Making Things: Generative Design*, released by Autodesk[3] on the future of generative design and manufacturing. Kowalski stated that "it's an entirely new way of doing things" and "everything that is designed will be affected".

Conti would like to design through a conversation with a computer. This philosophy is used already by software such as OpenAI's DALL-E, Midjourney, or Adobe Firefly, which can translate text prompts into images (Messer 2024). The software operates on concepts expressed in natural language and can show a corresponding result through artificial intelligence. According to Peckham et al. (2024), generative AI tools such as DALL-E can be used in the earlier stages of the creative process to generate concepts. In this regard, a discussion can be made about whether these generative AI tools are more creative than engineering software with generative design capabilities.

The idea of a prompt (i.e., the input statement introduced by the user in the software) is a critical element of generative AI models. Generating prompts is already an engineering area labeled as prompt engineering (or prompt design, prompt programming, or prompting) (Knoth et al. 2024). The prompt is of great relevance because the quality of the results depends on its definition.

Prompts ultimately challenge the way humans think about design. To illustrate, the generation of sketches is typical in design and art. The case of the Spanish painter Pablo Ruiz Picasso and his iconic painting *Guernica* is interesting. *Guernica* (Reina Sofia Museum, Spain) is one of the most significant paintings of the twentieth century. Its presentation at the 1937 *Exposition Internationale des Arts et Techniques dans la Vie Moderne* was held in Paris. From May 1 to June 4, Picasso made 45 sketches of various kinds for *Guernica* (Weisberg 2006). Undoubtedly, it is necessary to emphasize how specific the *Guernica* project was. However, generative AI can immediately provide artists and designers with numerous sketches, requiring them to invest considerable effort to achieve similar results.

Commercial software vendors have developed solutions based on traditional CAD frameworks for generative design. Thus, the design of the geometries follows the conventional methodologies used in CAD software, and the approach for developing generative design is a guided process that facilitates users' learning. In this context, the user must identify geometry (i.e., as starting or preserve geometry), apply load cases and constraints, indicate the materials and processes, etc. These operations must be done by selecting geometry and contextual menus and choosing from drop-down lists. As Peckham et al. (2024) stated, these commercial packages still need human designer guidance to drive the process. Moreover, it is important to note that generative design tools typically optimize designs for quantitative objectives. In this

---

[3] https://www.youtube.com/watch?v=E2SxqUvtpIk.

regard, designers must consider qualitative objectives such as aesthetics (Saadi et al. 2024).

Generative design offers a transformative perspective and a new interpretation of space. Thus, space is not occupied or left empty based on solly aesthetic criteria, but instead on a range of defined criteria. Efficiency guides the process; therefore, generative design allows designers to obtain more efficient designs, for instance, in terms of weight. It can generate a set of possible solutions, enabling the designer to select a definitive solution or continue working on one of the proposed options.

6.2 Exploring Differences Between Generative Design and Topology Optimization

The generative design concept is anchored to the structural optimization methods already introduced. As was mentioned, topology optimization is a process that helps designers develop more innovative designs than the ones resulting from shape and size optimization. Therefore, topological optimization relates more to generative design than the other two.

Topology optimization is a more established methodology as a search[4] in the Scopus database between 2000 and 2024 can indicate. The search included the terms "generative design" and "topology optimization". Moreover, "aerospace" was also combined with the two former terms. Figure 6.1 represents the number of results and shows a growing interest in both approaches. Topology optimization has a longer tradition and continues to receive more attention than generative design. There has been a growing interest in generative design over the past five to seven years. However, when referring to the aerospace sector, the number of results is still limited.

Sometimes, there is no clear distinction between topology optimization and generative design in popular and academic publications. These terms are occasionally used interchangeably, yet they represent distinct design concepts. Thus, it is essential to reflect on the different approaches they adopt.

Zhang et al. (2020) identified topology optimization as a reductionist and top-down strategy limiting innovations. As the authors stated: "the design space becomes increasingly limited as the design progresses and there is less opportunity for innovation". Despite the outcome produced by topology optimization being innovative and, thus, helping designers to create an unconventional design, the designer is given just one outcome (Srivastava and Kawakami 2023). In this sense, the designer is subjected to the phenomenon of design fixation, which can be defined as "blind adherence to a set of ideas or concepts limiting the output of conceptual design" (Jansson and Smith 1991; Wadinambiarachchi et al. 2024). However, the generative design process provides designers with various outcomes. Thus, the divergent creative process is stimulated.

---

[4] Data was obtained from the Scopus database on January 24, 2025. www.scopus.com.

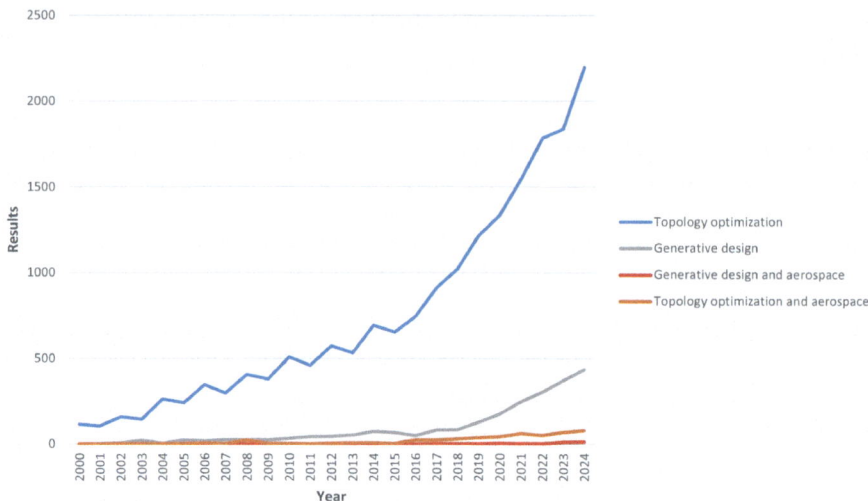

Fig. 6.1 Search in Scopus database from 2000 to 2024. Search within "article title, abstract, keywords"

Srivastava and Kawakami (2023) compared topology optimization and generative design, with the main points listed in Table 6.1 for illustration. Topology optimization requires a well-defined CAD and provides a single solution without analyzing function or manufacturability. In the case of generative design, there is no need to provide detailed starting geometries, and it offers several outcomes that consider not only manufacturability but also materials.

Despite the two approaches having notable differences, they can be used as complements. Jang et al. (2022) built on this idea to maximize the diversity of topology designs. These researchers proposed a reinforcement learning-based generative design process to create many solutions. The method was applied to an automobile wheel design. Gupta et al. (in press) have compared conventional design, generative design, and a fused alternative combining topology optimization and generative design for an RC plane landing gear.

6.3 The Workflow

Generative design comprises a set of activities that can be presented as a workflow. Designers must carefully analyze their problem and define the key elements and limits that satisfy the functional requirements of the part or assembly to be developed. Then, generative design is a tool for creating a large set of concepts when the number of design variables is large.

6.3 The Workflow

Table 6.1 Comparison of topology optimization and generative design (Srivastava and Kawakami 2023)

Topology optimization	Generative design
• The manufacturing and functional needs are settled after the outcome is generated. It does not routinely consider manufacturability	• It effectively allows for the definition of a design problem by stating the goals and requirements for both manufacturing and functional needs
• To meet the specified performance criteria, optimization uses a single or multi-material topology, assigning materials to different areas of the design domain	• It enables the selection of different materials to consider while setting up the study, which increases the range of possible outcomes
• A thoroughly defined CAD model of the original geometry is necessary for the process	• Based on loading conditions and constraints, design space can be created for generative design
• The method relies on a mathematical method to subtract or add material by stress	• To provide results, the procedure uses topology optimization, machine learning, artificial intelligence, and several strategies
• It provides only one solution and is suitable for conceptualization	• It produces a variety of outcomes, notably some of which are ready for manufacturing with some post-processing

The workflow for the generative design includes the following steps (Buonamici et al. 2020; Srivastava and Kawakami 2023)[5]:

- Establishing the objective. The designer must establish the goals for the analysis, typically aimed at minimizing bulk or maximizing stiffness.
- Identifying the geometries. The designer specifies the volumes that must stay empty (obstacle geometry), the regions that should be preserved during optimization (preserve regions), and, optionally, a starting volume.
- Defining the load cases. Generative design uses forces, pressures, bearing loads, and gravity in the analysis. Moreover, three types of constraints can be established: fixed, pinned, and frictionless. It is not possible to analyze dynamic conditions.
- Selecting the manufacturing process. Various processes are available, including additive manufacturing, 5-axis milling, 3-axis milling, and an unrestricted option.
- Selecting materials. It is possible to select up to ten materials.
- Performing the input check and calculation. The software verifies that the necessary data is provided. When all the requirements are fulfilled, the software performs the calculations.
- Visualizing the results. The designer is provided with a set of outcomes based on the prior decisions.
- Performing an exploratory analysis. The software has a dedicated environment that aids designers in organizing their exploration of the outcomes.
- Selecting and exporting. The user chooses a design from the provided outcomes and exports it. The CAD geometry is in the Fusion 360 modeling environment.

---

[5] The workflow is based on the generative design module in Autodesk Fusion 360.

- Post-processing. The designer can fix any flaws that are frequently found in complex designs.
- Validating. If no discrepancies are detected, the post-processed model can be compared with the original model to validate the final design, ensuring consistency. This stage can be done using finite element analysis.

Although the workflow may vary depending on the chosen software, certain common elements, such as geometry, will appear regardless of the software solution used in the design. These common elements include the following, based on the documentation by Autodesk[6]:

- Starting shape (optional): assigned to a body to optimize an existing design or influence the shape of the generated design. It is assigned to only one body; the software will create more outcome variation if no starting shape exists. The body should be in contact with all preserve geometry bodies.
- Preserve geometry: assigned to bodies integral to the final design shape, whose geometry remains unchanged throughout. At least one preserve geometry body is needed to run the simulation.
- Obstacle geometry: assigned to bodies to represent spaces that must be avoided in the design.

As generative design is a methodology that freely uses the entire design space, the designer must indicate any non-design spaces. These spaces can be identified as obstacle geometry (red color) and preserve geometry or fixed volume (green color). Figure 6.2 shows an example of generative design applied to an aircraft bearing bracket (Pilagatti et al. 2023). In the figure, it is possible to appreciate the obstacle geometries: ball bearing aircraft connector, sensor system (rotating shaft, ball bearings, motor, and counter mass) and the electronic components. Moreover, readers can see the fixed volume: coupling volume with the aircraft, coupling volume with rolling bearings, and volume for supporting electronic components. Finally, the figure does not have a starting shape, which should appear in yellow.

6.4 The Biological Inspiration

Many designs created using generative design methodologies resemble biological forms. Biology has long inspired designs; birds inspired Leonardo da Vinci to design his flying machines. The generative design results have captured the attention of designers, who labeled them innovative and novel. One reason that may explain this is that designers and engineers still require a biological background (Willocx et al. 2021).

---

[6] https://help.autodesk.com/view/fusion360/ENU/?guid=GD-DESIGN-SPACE.

6.4 The Biological Inspiration

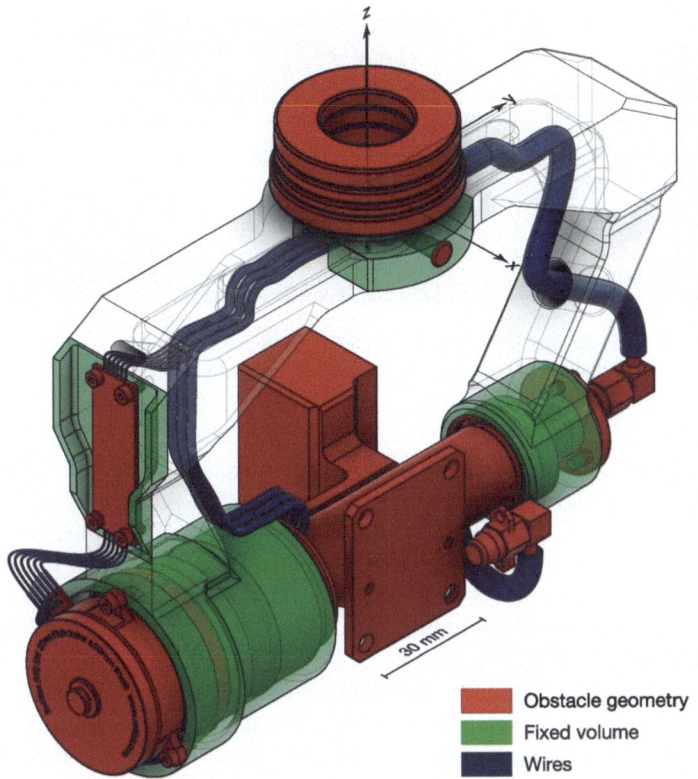

Fig. 6.2 Detail of the non-design spaces for analyzing an aircraft bearing bracket (Pilagatti et al. 2023)

Focusing attention on biology is simply because it works. As Malshe et al. (2023) stated, "biology, through millions of years of evolution, has met these acute requirements under severe resource and environmental constraints". In this sense, biology can inspire manufacturing because it also has important constraints, such as using raw materials, labor, and other resources. Therefore, there is an analogy between biological and engineering problems. Specifically, bio-inspired designs can be used in many aerospace applications (Rose et al. 2021).

According to van Houten et al. (2021), bio-inspiration can help designers in two ways: "as a trigger for searching and implementing a solution for an engineering problem or as a trigger by using observation in the bio-space to be converted or adapted to the technical space". One example of an invention based on bio-inspired design is Velcro®. George de Mestral created the product after observing burdock seeds stuck in the dog's fur (Wang and Deng 2022).

Fig. 6.3 Bio-inspired design: (left) Beehive, a standard honeycomb structure in nature; (right) deck components: the upper shell and the assembled bottom shell with honeycomb core (Namvar et al. 2023)

The scientific community has been paying increasing attention to the term biomimetic, particularly since 2005. Biomimicry is "the practice of making technological and industrial design copy natural processes".[7] Biological systems are complex; thus, copying and obtaining industrial replicas was difficult with conventional manufacturing technology. However, the new additive manufacturing processes connect well with biomimicry and allow faithful replications of biological models (du Plessis et al. 2019).

Understanding biology is an attractive and novel background that can help designers and engineers develop more efficient and innovative designs (Malshe et al. 2023). For instance, having a background in biological materials and structures is essential. There are numerous lightweight biological structures that provide high energy absorption capacity that may serve as inspiration (Tee et al. 2021; Mohammadi et al. 2023), such as beehive structures (see Fig. 6.3, which illustrates how nature's honeycomb structure inspires modern aerospace engineering designs).

Cellular materials have attracted significant attention in recent years. They can be found in nature (trabecular bones, fungi, and honeycomb structures). Some reported advantages of cellular materials include high strength, stiffness, impact resistance, thermal and acoustic properties, and, in general, attractive multi-functional and versatile behavior. Current applications include lightweight aerospace and automotive engineering components, thermal and sound insulation, biomedical and tissue engineering, energy, and impact absorption (Colamartino et al. 2023). Cellular materials can be manufactured using well-defined periodic geometries, and as a result, their mechanical properties are expected to be controllable (Benedetti et al. 2021).

Biomimetics can be applied in the design process. In this sense, Gralow et al. (2020) presented a four-stage methodology for developing a biomimetics design process. These stages include problem definition, analogy search and preselection, analysis and abstraction of the biological model, and transfer and product development. A couple of examples of the application of the method were discussed, specifically, a bracket from the Airbus A380 series that is a part of the fixation of

---

[7] https://dictionary.cambridge.org/dictionary/english/biomimicry.

the flight crew rest compartment with the primary aircraft hull and a bionic partition that separates the passenger cabin and the galley.

6.5 Generative Design Examples

The advance of additive manufacturing allows researchers to rethink design and manufacturing processes. Joints for treelike column structures (Wang et al. 2021a) are traditionally manufactured and welded to the rest of the structure. However, additive manufacturing combined with generative design allows the creation of more complex geometries without welding operations.

Architecture is a field in which aesthetics and innovation often drive the design process. Therefore, generative design is an ideal methodology for developing innovative projects (Teixeira et al. 2023). To illustrate, Autodesk used the methodology to design its Toronto office[8] (Walmsley 2022). The project pushes the boundaries of generative design by evaluating 10,000 design options. The sheer volume of possible solutions is unmanageable when considering the traditional design process of any architectural firm worldwide.

Multiple examples can be cited to help readers understand the importance of the methodology far beyond architecture. By way of illustration, generative design has been explored in diverse applications, including:

- 3D-printed hands-free door handles for reducing contagion risk in public buildings (Crescenzio et al. 2022).
- 3D concept wheels (Yoo et al. 2021).
- Bicycle frames (Ondriga et al. 2023; Vigil-Fernandez et al. 2024).
- Front axle wheel hub (Szabó 2023).
- Furniture (Lana et al. 2023).
- Heat exchangers (Li et al. 2022).
- Hydraulic manifold components (Wang et al. 2021b).
- Maxillo-facial reconstructive surgery (Filippi et al. 2021).

In aerospace, there are already iconic examples such as the project carried out by Airbus AP Works and Autodesk to develop a bionic cabin partition (Lathabai 2018; Carou 2021). Next, some noteworthy applications of the methodology that show the potential for new designs in aerospace will be presented. Most of these were developed to leverage additive manufacturing as the manufacturing process and, in most cases, were created using Fusion 360 by Autodesk.

Generative design software has experienced a vital impulse that helps develop applications in recent years. Several software vendors assert their packages offer

---

[8] https://www.architectmagazine.com/project-gallery/autodesk-mars-office_o.

generative design capabilities in their commercial offerings.[9] Some of the most notable vendors and their solutions include Fusion 360 (Autodesk, USA, www.autodesk.com), CogniCAD (Carbon®, USA), www.carbon3d.com), Creo 7.0 (PTC, USA, www.ptc.com), MSC Apex (Hexagon, USA, www.hexagon.com), Solid Edge (Siemens, Germany, www.siemens.com), NX (Siemens, Germany, www.siemens.com), and CATIA V6 (Dassault Systèmes, France, www.3ds.com) (Junk and Burkart 2021).

6.5.1 Aircraft Bearing Bracket

Pilagatti et al. (2023) employed a generative design approach to create an aircraft bearing bracket that supports sensors mounted on a rotating shaft, facilitated by rolling bearings. The part has specific requirements to fulfill. For instance, it must guarantee 1.5 mm of maximum deformation, 2 as a minimum safety factor, and Ø0.6 mm of position tolerance between the axes of the two-bearing housing in the deformed configuration.

The original bracket has dimensions of 162 mm × 116 mm × 47 mm and is made of UNI EN AW-7075 (Ergal) alloy. The *RTCA DO-160G-Environmental Conditions and Test Procedures for Airborne Equipment* standard establishes the conditions to be tested. Therefore, the bracket must respond adequately to six possible acceleration cases. The design included a 522 g mass sensor connected to a 19 mm diameter Ergal shaft. Regarding the load cases, the authors analyzed twelve load cases due to the combination of two different torques and six possible acceleration cases.

The bracket design is shown in Fig. 6.2 with the non-design spaces. Moreover, it is necessary to establish the objectives for conducting the generative design process. The aim was to reduce the mass by constraining the maximum permissible von Mises stress. The materials to be used are the AlSi10Mg and Ti6Al4V alloys because the original material was not available for additive manufacturing (Laser-Powder Bed Fusion, L-PBF) at the time of the research. The L-PBF process enables the creation of complex geometries but still has some manufacturing limitations, such as the maximum permissible angle that can produced without supporting cantilevered surfaces (45° for AlSi10Mg and 35° for Ti6Al4V).

The software used to perform the generative design was Fusion 360 by Autodesk. In Fig. 6.4, the 16 solutions proposed are visible. The software ranks the solutions based on material usage (mass). Thus, aluminum solutions were ranked from 1 to 8. Moreover, the unconstrained solutions allowed manufacturing parts with the lowest mass. The GD(13) output was the sole titanium output that enabled weight reduction.

The authors analyzed the part's manufacturing using Autodesk Netfabb Premium software, considering the volume of material needed for the part and support, and the

---

[9] Whether the claim is fully accurate or not is not discussed in the present study. A discussion on the real capabilities of some design software can be found in *A Study on the Application of Generative Design in Aircraft Conceptualization* by Santiago Mañé Ubalde (Ubalde, 2023).

6.5 Generative Design Examples

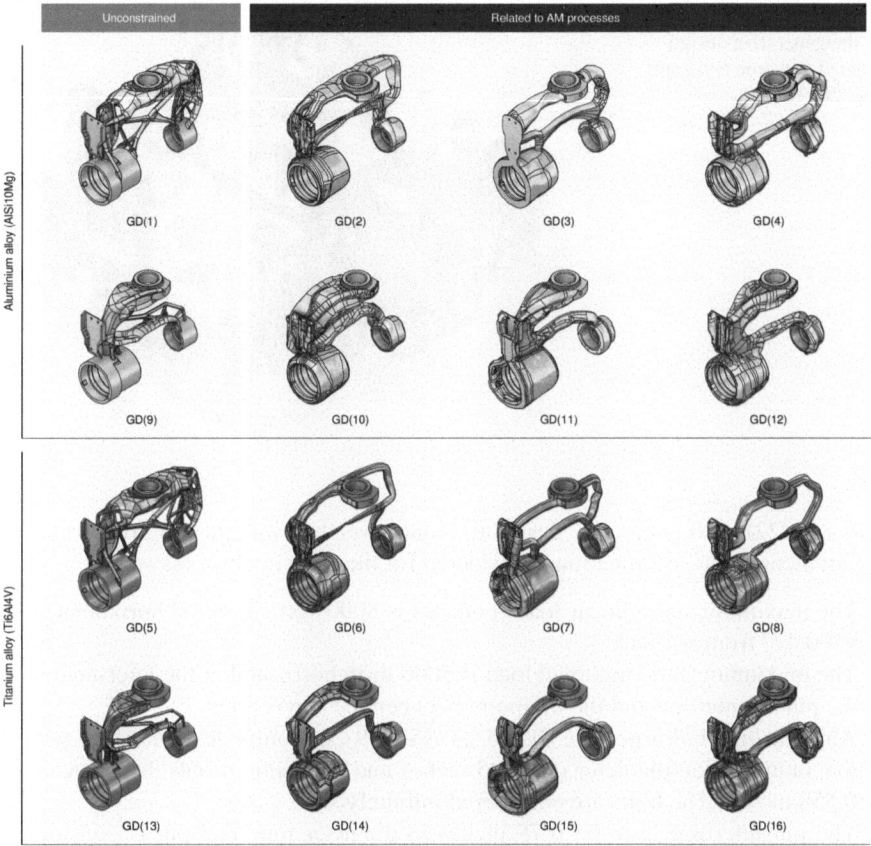

Fig. 6.4 Outputs of the generative design process for aluminum and titanium alloys for both unconstrained and constrained additive manufacturing (Pilagatti et al. 2023)

building time of a single component. The L-PBF technique requires using supports that help anchor the part to the building platform, dissipate the excess heat, and stabilize features that exceed the allowable overhang angle. The EOSM 400-4 system enabled manufacturing up to 12 parts per job (Fig. 6.5). A layer height of 80 μm was used. The build time per job was 182.9 h.

6.5.2 Aircraft Engine Loading Brackets

The engine loading brackets support the engine's weight during operations without breaking or warping. These brackets remain installed on the engine. Botyarov and

Fig. 6.5 Building platform for the generative design selected outcome (Pilagatti et al. 2023)

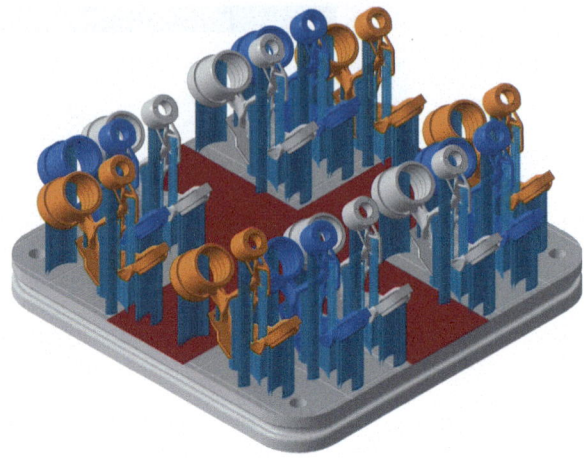

Miller (2022) used generative design to reduce weight while meeting performance requirements. The specifications considered for the redesign process were:

- The maximum static linear load (pounds) is 8000 vertical, 8500 horizontal, and 9500 42° from vertical.
- The maximum static torsional load is 5000 lb-in horizontal at the intersection of the pin's centerline and the midpoint between the clevis arms.
- Any machine bolt interface (0.375-24 AS3239-26) comprises a nut face with a maximum inside diameter of 0.405 inches and a minimum outside diameter of 0.558 inches. The bolts are considered infinitely stiff.
- The pin interface is to be 0.75 inches in diameter pin. The pin is considered infinitely stiff.

The authors adjusted the safety factor to obtain a large number of solutions (i.e., 1000 solutions). The example helps illustrate the power of the methodology. However, designers need help finding optimal solutions among numerous outcomes when considering multiple factors.

To find a suitable solution, the authors proposed a clustering method. This approach progressively reduces the number of solutions in steps (1000, 500, 250, 100, 50, 25, 10) while maintaining those whose results are close. The process was performed using the R software. Clustering was done by combining the Gower distance matrix and partitioning around medoids in an iterative process. Six iterations were performed to reach the final group of ten solutions. The analysis was based on two nominal (material, manufacturing method) and three continuous (volume, mass, max von Mises stress) properties.

It is interesting to indicate the materials and manufacturing processes that were analyzed. In this regard, the following were considered:

- Materials: aluminum 7175 T73 0 hot formed, aluminum AlSi10Mg, cobalt chrome, Inconel 625, Inconel 718, Inconel 718 Plus, Iron (cast), Iron (ductile),

6.5 Generative Design Examples

Iron (malleable), stainless steel 17-4 PH, stainless steel AISI 304, stainless steel AISI 440C (welded), stainless steel 440C, and titanium 6Al-4V.
- Manufacturing: 2-axis cutting, 2.5-axis milling, 3-axis milling, 5-axis milling, additive, die casting, and unrestricted.

The process of reducing the initial space of 1000 to 10 solutions leads to the solutions shown in Fig. 6.6. Eight out of the fourteen materials and five out of the seven manufacturing processes are included in the solutions. When plotting the solutions against the studied variables, the authors found no overrepresented areas within any region of the plots.

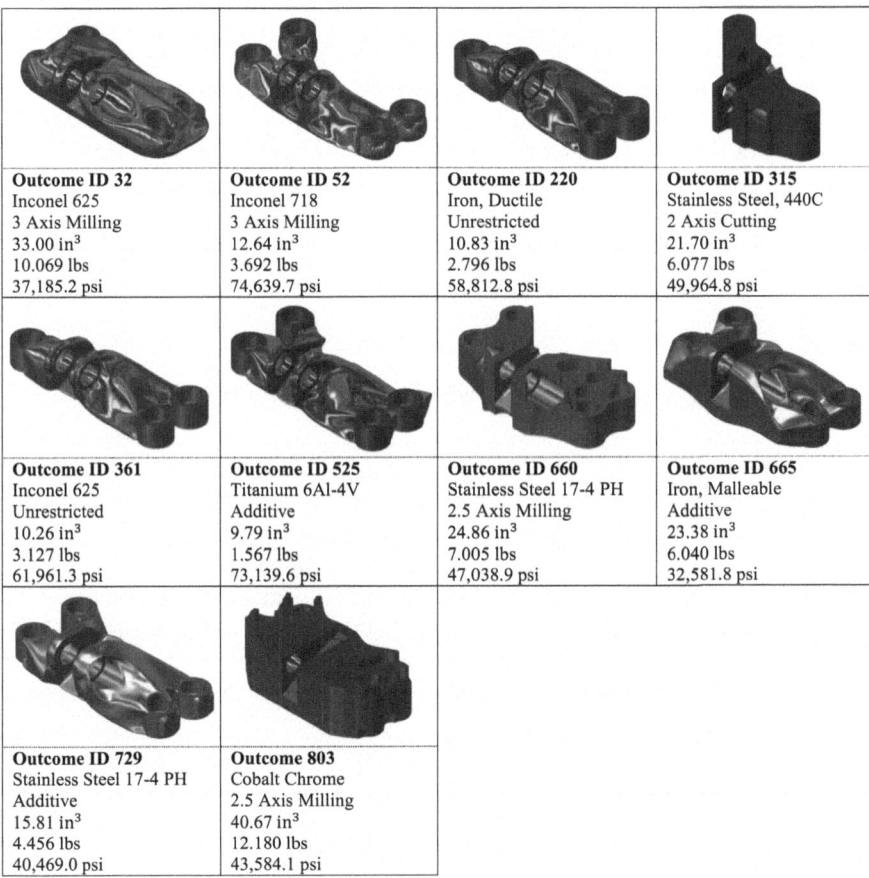

Fig. 6.6 Final generative design solutions (Botyarov and Miller 2022)

6.5.3 Bionic Partition

The bionic partition project by Airbus is a notable example of the potential of both generative design and additive manufacturing. The partition divides the passenger seating area and the plane's galley. Some seating configurations support the jumpseats used by cabin crews. The project is a collaborative effort involving Airbus, Autodesk (with experts from The Living, an Autodesk Studio), and printing experts from APWorks (Carou 2021; Lathabai 2018; Walmsley 2022).[10]

The bionic partition used the Scalmalloy® material. This lightweight material with outstanding mechanical properties was specifically designed for additive manufacturing. This is reportedly the first large-scale application of the material within an aircraft component. At the project's core, generative design allows for finding an optimized geometry from thousands of design alternatives that meet the established specific goals and constraints. The generative design results resemble biology. Specifically, the algorithm used was based on the adaptive networks of slime mold.[11]

The generative design aimed to minimize both weight and displacement under structural load. The resulting cabin partition satisfied the structural requirements, achieving a 45% weight reduction (30 kg) compared to current designs. Airbus estimates the new design can save up to 465,000 metric tons of CO_2 emissions annually.

6.5.4 Landing Gear of an Ultralight Single-Seater Aircraft

The generative design was used by Pilagatti et al. (2022) to study an assembly part of the tail landing gear of an ultralight single-seater aircraft, the Zigolo MG12, by Aviad. The assembly includes several parts joined together using bolts.

The authors performed the generative design simulation using Autodesk Fusion 360. As a constraint, the required safety factor was 1.5. The objective was to reduce the original mass. The selected process was L-PBF. Because the original Al2020 alloy was considered inadequate for additive manufacturing, the authors used the EOS AlSi10Mg alloy, which offers a high strength-to-weight ratio, good thermal conductivity, and corrosion resistance.

The authors analyzed the load cases of landing and right and left steering, with values of 1069 and 854 N, respectively. The geometry included the starting, obstacle, and preserve geometry, as shown in Fig. 6.7.

Fusion 360 provides a set of outcomes based on the specified criteria. Thus, one outcome was developed based on an unrestricted manufacturing process, and six were developed for additive manufacturing. The authors created a ranking score to analyze the outcomes based on objective criteria, using the component's weight, cost,

---

[10] https://www.airbus.com/en/newsroom/news/2016-03-pioneering-bionic-3d-printing.

[11] https://www.architectmagazine.com/technology/the-living-and-autodesk-apply-bionic-design-to-an-airbus-320-partition_o.

6.5 Generative Design Examples

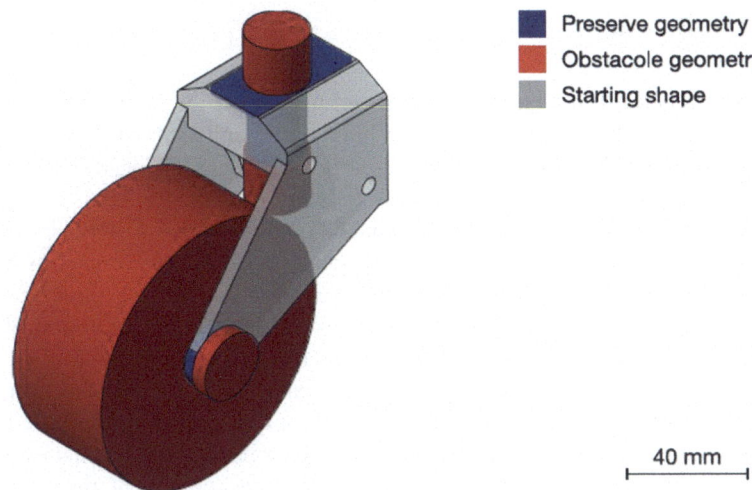

Fig. 6.7 3D model to use in the generative design (Pilagatti et al. 2022)

and maximum displacement. Equation 6.1 presents the way to calculate the score for the i-th outcome:

$$\text{Score}(i) = \frac{x_{\max} - x_i}{x_{\max} - x_{\min}}(r - 1) + 1 \tag{6.1}$$

where x_i is the value of the parameter, and $x_{\max}$ and $x_{\min}$ are the maximum and minimum values, respectively. The maximum score of the criterion is r (6). Weights for the three factors were 1 for mass and 0.5 for cost and displacement. In Table 6.2, it is possible to see all the results and the total weighted score for each outcome. The best outcome based on Eq. 6.1 was outcome #5. However, the authors edited the outcome as it was not entirely symmetrical. Thus, the part was divided in half and mirrored to obtain a symmetric result. The generative design was run twice to reduce the mass from 72 to 67 g. This represents a significant reduction from the 148 g of the original assembly.

Table 6.2 AlSi10Mg outcome selection for a productive batch of 50 pieces

Outcome	Mass (g)	Displacement (mm)	Cost (€)	Total weighted score
#5	72	0.86	114.54	9.80
#3	76	0.95	90.47	9.52
#6	72	1.02	113.71	9.03
#2	72	1.07	114.54	8.69
#1	75	1.33	117.03	5.58
#4	82	0.99	117.86	3.81

6.5.5 GE Bracket

The General Electric (GE) bracket is designed to support the load acting on a jet engine. Marconi et al. (2022) performed an initial structural analysis through Ansys Workbench 2021 R2. The material used was structural steel of a density of 7850 kg/m^3, Young's modulus of 200 GPa, and Poisson's ratio of 0.3. The results include a mass of 3.636 kg, deformation of 0.089 mm, and equivalent stress of 423.91 MPa.

The authors compared the use of different types of design software, such as Ansys Workbench 2021 R2 (topology optimization), Autodesk Fusion 360 2022 (topology optimization and generative design), and Siemens NX 12.0 (topology optimization). The results of the software packages are listed in Table 6.3.

Table 6.3 shows how all four solutions significantly reduce the weight from the original design. However, deformations increased for the four cases, and the equivalent stress increased for the Fusion 360 solutions. These results indicate that while the software solutions can effectively reduce the weight of the component, they may also lead to increased deformations and stresses, which should be carefully considered in the design process.

Figure 6.8 presents images of the four solutions. Upon comparison, it is clear that Fusion 360's generative design solution is the least conventional of the four. It is less intuitive and, therefore, unlikely to be proposed by an industrial designer.

Table 6.3 Comparison of the different GE bracket solutions

Software	Mass (kg)	Deformation (mm)	Equivalent stress (MPa)
Ansys Workbench	1.876 (− 48.4%)	0.107 (+ 20.2%)	417.9 (− 1.4%)
Fusion 360 TO	1.861 (− 48.8%)	0.113 (+ 27.0%)	542.7 (+ 28.0%)
Siemens NX	1.811 (− 50.2%)	0.131 (+ 47.2%)	368.5 (− 13.1%)
Fusion 360 GD	0.947 (− 74.0%)	0.413 (+ 364.0%)	969.5 (+ 128.7%)

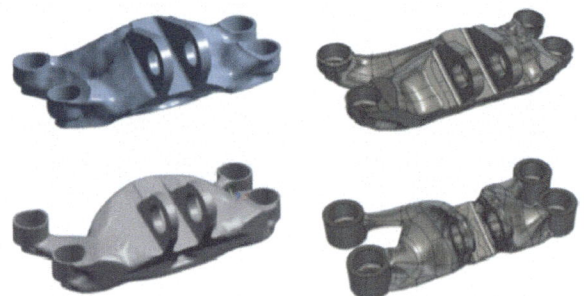

Fig. 6.8 GE bracket: (up_left) Ansys Workbench, (up_right) Fusion 360 topology optimization, (down_left) Siemens NX, (down_right) Fusion 360 generative design (Marconi et al. 2022)

6.5 Generative Design Examples

6.5.6 Drone Frames

In recent years, drones have been used in various sectors to perform diverse tasks (Carou et al. 2024). Weight is a critical consideration when designing a drone, as it is directly related to the generation of lift required to ensure the drone's flight (Orgeira-Crespo and García-Luis 2024). Thus, special care must be taken during the design phase and the selection of the drone's components.

6.5.6.1 Simplified Drone Frame

Marconi et al. (2022) used the same methodology presented for the GE bracket to analyze a simplified drone frame. The frame supports four electric motors, an electronic board for remote control, and a rechargeable battery pack. The initial structural analysis, conducted using Ansys Workbench 2021 R2, employed the same structural steel as the GE bracket. The results include a mass of 0.153 kg, deformation of 0.0047 mm, and equivalent stress of 2.45 MPa.

The comparison was made using Ansys Workbench 2021 R2, Autodesk Fusion 360 2022, and Siemens NX 12.0. Figure 6.9 shows the geometry of the four analyses, and Table 6.4 lists the results.

These studies can attain considerable mass reductions. The deformations increased from the original solution, except for the Fusion 360 generative design solution. This solution differs significantly from the others, not including a hollow space in the middle. Regarding the equivalent stress, the values are higher than those of the initial solution. Remarkably, the values for Fusion 360 are more than three times higher.

Fig. 6.9 Drone frame: (up_left) Ansys Workbench, (up_right) Fusion 360 TO, (down_left) Siemens, (down_right) Fusion 360 GD (Marconi et al. 2022)

Table 6.4 Comparison of the different drone frame solutions

Software tool	Mass (kg)	Deformation (mm)	Equivalent stress (MPa)
Ansys Workbench	0.091 (− 40.5%)	0.0057 (+ 21.3%)	2.99 (+ 22.0%)
Fusion 360 TO	0.084 (− 45.1%)	0.0089 (+ 89.4%)	7.92 (+ 223.3%)
Siemens NX	0.076 (− 50.3%)	0.0061 (+ 29.8%)	4.27 (+ 74.3%)
Fusion 360 GD	0.045 (− 70.6%)	0.0027 (− 42.6%)	7.72 (+ 215.1%)

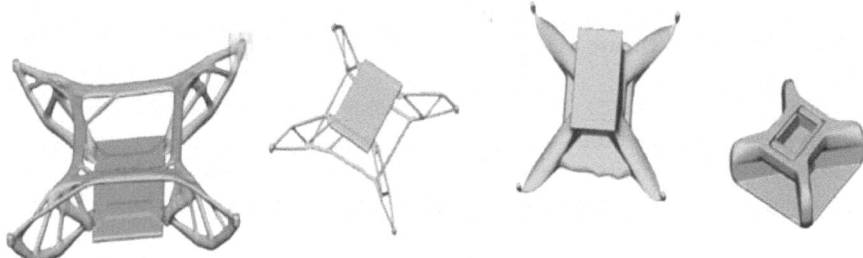

Fig. 6.10 Selected outcomes for the frame structure (Yadav et al. 2021)

6.5.6.2 A Frame of a Quadcopter Drone

Generative design is an adequate tool for exploring design ideas to develop optimized frame structures. Yadav et al. (2021) used the methodology to study the frame structure of a quadcopter. The body frame to analyze in the case study includes the base and the arms. Moreover, two identical portions make up the upper and lower foundation frames.

The generative design process enabled the creation of various designs meeting the requirements. The authors selected four outcomes to be further analyzed, as shown in Fig. 6.10. The generative design approach allowed researchers to create a new design that helps reduce weight and improve load-bearing capacity.

6.5.7 Wing Ribs of a Remote-Controlled Plane

D'mello et al. (2020) conducted a generative design study to improve wing ribs. The ribs are elements that provide the structure with the required stiffness. The ribs that the authors studied were those of the plane Chico-1, designed by Team Albatross for the SAE Aero Design competition. The study was performed using Ansys Fluent and Autodesk Fusion 360.

The wing details include a total wing span of 3.2 m and 0.146 m inside the wing. The study was done to analyze one half of the wing (the outside part), so a length of 1.527 m, including 11 ribs. Moreover, the width of the fuselage was 0.146 m. The profile airfoil was Selig 1223. The design included a hollow square aluminum spar of outer diameter 0.012 m, passing through each wing rib at a quarter of the distance of the mean aerodynamic chord from the leading edge of the rib. The hollow spar was fitted into another hollow aluminum spar. The total weight of the part of the wing being analyzed was 94.38 g.

An initial computational fluid dynamics (CFD) study in Ansys Fluent allowed the authors to obtain the loads applied to each rib. The second and third ribs exhibited the highest lift force (6.83 N). In addition, the load, dependent on the rib's mass, was applied at the rib's center of gravity. As a constraint, a square was designed as the spar hole, and the four walls were fixed.

The authors selected additive manufacturing, 3-axis milling, and the unrestricted alternative as manufacturing options. The materials were acrylonitrile butadiene styrene (ABS) and polylactic acid (PLA). The safety factor was fixed at 1.2, and the goal was to minimize mass. By reducing the mass of the rib, the weight of the wing was also reduced. Thus, the payload carrying capacity can be increased for the same lift force.

The generative design results included ten outcomes consisting of completed and converged solutions. The authors selected an outcome for additive manufacturing in Y+ with a mass of 15 g when printing in PLA and a safety factor of 1.61 (exceeding the established 1.2 at the outset). The maximum displacement was 1.69 mm.

6.5.8 Flapping Wings

Jiang et al. (2023) conducted a generative design study on the exploration of flapping wing shapes in flapping wing air vehicles (FWAV). The authors based their work on the bio-inspired generative design framework comprising three stages: dataset building, generator modeling, and design evaluation.

The case study relies on using training datasets with biological knowledge. The authors used web crawlers, such as Google Images and the Natural History Museum's wing database. Three types of birds, falcons, hummingbirds, and sparrows, were explicitly analyzed. The generative adversarial network (GAN) model was built to transfer the information of natural insect wings to the developed wing models. The generated wings were expected to produce maximized lift for various flight conditions: hovering, high speed, and low speed. The best lift performance was attained as follows: hummingbird-shaped wings (hovering conditions), falcon-shaped wings (high-speed flight), and sparrow-shaped wings (low-speed flight). The authors noted that lift performance closely corresponds to biological knowledge.

References

Benedetti, M., A. du Plessis, R.O. Ritchie, M. Dallago, N. Razavi, and F. Berto. 2021. Architected cellular materials: A review on their mechanical properties towards fatigue-tolerant design and fabrication. *Materials Science and Engineering R* 144: 100606.

Botyarov, Michael, and Erika E. Miller. 2022. Partitioning around medoids as a systematic approach to generative design solution space reduction. *Results in Engineering* 15: 100544.

Buonamici, Francesco, Monica Carfagni, Rocco Furferi, Yary Volpe, and Lapo Governi. 2020. Generative design: An explorative study. *Computer-Aided Design and Applications* 18: 144–155.

Carou, Diego. 2021. *Aerospace and digitalization. A transformation through key industry 4.0 technologies*, ed. J. P. Davim. Springer Nature.

Carou, Diego, Antonio Sartal, and J. Paulo Davim. *Applying drones to current societal and industrial challenges*. Springer.

Colamartino, Ivan, Marco Anghileri, and Marco Boniardi. 2023. Investigation of the compressive properties of three-dimensional Voronoi reticula. *International Journal of Solids and Structures* 284: 112501.

D'mello, Shawn Joseph, S. Renold Elsen, and J. Ronald Aseer. 2020. Generative design study of a remote-controlled plane's wing ribs. *AIP Conference Proceedings* 2283: 020046.

De Crescenzio, F., M. Fantini, and E. Asllani. 2022. Generative design of 3D printed hands-free door handles for reduction of contagion risk in public buildings. *International Journal on Interactive Design and Manufacturing (IJIDeM)* 16: 253–261.

du Plessis, Anton, Chris Broeckhoven, Ina Yadroitsava, Igor Yadroitsev, Clive H. Hands, Ravi Kunju, and Dhruv Bhate. 2019. Beautiful and functional: A review of biomimetic design in additive manufacturing. *Additive Manufacturing* 27: 408–427.

Filippi, Stefano, Massimo Robiony, Alessandro Tel, and Francisco Daniel Montoya Buteler. 2021. Limitations of using generative design for maxillo-facial reconstructive surgery applications. In: *Design Tools and Methods in Industrial Engineering II. Proceedings of the Second International Conference on Design Tools and Methods in Industrial Engineering, ADM 2021*, September 9–10, 2021, Rome, Italy, ed. Caterina Rizzi, Francesca Campana, Michele Bici, Francesco Gherardini, Tommaso Ingrassia, and Paolo Cicconi.

Gralow, Melanie, Felix Weigand, Dirk Herzog, Tim Wischeropp, and Claus Emmelmann. 2020. Biomimetic design and laser additive manufacturing—A perfect symbiosis? *Journal of Laser Applications* 32: 021201.

Gupta, Aditya, Vedant Soni, Dhaval Shah, Absar Lakdawala. IN PRESS. Generative design of main landing gear for a remote-controlled aircraft. *Materials Today: Proceedings*.

HBR. 2018. *The next wave of intelligent design automation*. Harvard Business Review Analytic Services.

Jang, Seowoo, Soyoung Yoo, and Namwoo Kang. 2022. Generative design by reinforcement learning: Enhancing the diversity of topology optimization designs. *Computer-Aided Design* 146: 103225.

Jansson, David G., and Steven M. Smith. 1991. Design fixation. *Design Studies* 12 (1): 3–11.

Jiang, Zhoumingju, Yongsheng Ma, and Yi Xiong. 2023. Bio-inspired generative design for engineering products: A case study for flapping wing shape exploration. *Advanced Engineering Informatics* 58: 102240.

Junk, Stefan, Lukas Burkart. 2021. Comparison of CAD systems for generative design for use with additive manufacturing. 2. 31st CIRP Design Conference 2021 (CIRP Design 2021). *Procedia CIRP* 100: 577–58. 10.1016/j.procir.2021.05.126

Kallioras, Nikos Ath., and Nikos D. Lagaros. 2020. DzAIN: Deep learning based generative design. *Procedia Manufacturing* 44: 591–598.

Knoth, Nils, Antonia Tolzin, Andreas Janson, and Jan Marco Leimeister. 2024. AI literacy and its implications for prompt engineering strategies. *Computers and Education: Artificial Intelligence* 6: 100225.

References

Lana, Jarža, Prekrat Silvana, and Pervan Stjepan. Sustainable furniture modelling by using generative design. In: *7th International Scientific Conference A.L.I.C.E. 2023 Consumerism and Sustainability*.

Lathabai, Sri. 2018. Additive manufacturing of aluminium-based alloys and composites. In *Fundamentals of aluminium metallurgy. Recent advances*, 47–92. Woodhead Publishing Series in Metals and Surface Engineering.

Li, Ning, Jean-Michel. Hugo, Damien Serret, Yicha Zhang, and Samuel Gomes. 2022. A two-step parametric generative method for heat exchangers design in additive manufacturing. *Procedia CIRP* 109: 508–512.

Malshe, Ajay P., Salil Bapat, Kamlakar P. Rajurkar, Ang Liu, and Jean-Marc. Linares. 2023. Exploring the intersection of biology and design for product innovations. *CIRP Annals-Manufacturing Technology* 72: 569–592.

Marconi, Marco, Alessio Zanini, Claudio Favi, and Marco Mandolini. 2022. Design for additive manufacturing tools: are they an effective support for designers? In: *Advances on Mechanics, Design Engineering and Manufacturing IV. Proceedings of the International Joint Conference on Mechanics, Design Engineering and Advanced Manufacturing, JCM 2022*, June 1–3, 2022, Ischia, Italy, ed. Salvatore Gerbino, Antonio Lanzotti, Massimo Martorelli, Ramón Mirálbes Buil, Caterina Rizzi, and Lionel Roucoules. Springer.

Messer, Uwe. 2024. Co-creating art with generative artificial intelligence: Implications for artworks and artists. *Computers in Human Behavior: Artificial Humans* 2 (1): 100056.

Mohammadi, Hossein, Zaini Ahmad, Michal Petrů, Saiful Amri Mazlan, Mohd Aidy Faizal Johari, Hossein Hatami, Seyed Saeid Rahimian Koloor. 2023. An insight from nature: Honeycomb pattern in advanced structural design for impact energy absorption. *Journal of Materials Research and Technology* 22: 2862–2887.

Namvar, Naser, Ilya Moloukzadeh,·Ali Zolfagharian, Frédéric Demoly, and Mahdi Bodaghi. 2023. Bio inspired design, modeling, and 3D printing of lattice based scale model scooter decks. *The International Journal of Advanced Manufacturing Technology* 126: 2887–2903.

Ondriga, Jozef, Lucia Piatrikova, Jozef Jenis, and Slavomir Hrcek. 2023. Generation of bicycle frame image designs using DCGAN network. *Transportation Research Procedia* 74: 688–693.

Orgeira-Crespo, Pedro, and Uxía García-Luis. 2024. Brief introduction to unmanned aerial systems. In: *Applying drones to current societal and industrial challenges*, ed. D. Carou, A. Sartal, and J. Paulo Davim. Springer.

Oyesola, Moses O., Khumbulani Mpofu, Ntombi R. Mathe, and Ilesanmi Daniyan. 2019. Development of an integrated design methodology model for quality and throughput of Additive Manufacturing processes. *Procedia CIRP* 84: 688–693.

Peckham, Owen Rahmat, Ben Hicks, and Mark Goudswaard. 2024. Are generative design tools creative? A characterisation of tools throughout the design process. In *International Design Conference—Design*, 683–692.

Pilagatti, Adriano Nicola, Giuseppe Vecchi, Eleonora Atzeni, Luca Iuliano, and Alessandro Salmi. 2022. Generative design and new designers' role in the manufacturing industry. *Procedia CIRP* 112: 364–369.

Pilagatti, Adriano Nicola, Eleonora Atzeni, and Alessandro Salmi. 2023. Exploiting the generative design potential to select the best conceptual design of an aerospace component to be produced by additive manufacturing. *The International Journal of Advanced Manufacturing Technology* 126: 5597–5612.

Rose, J. Bruce Ralphin, S. Ganesh Natarajan, and V. T. Gopinathan. Biomimetic flow control techniques for aerospace applications: a comprehensive review. *Reviews in Environmental Science and Bio/Technology* 20: 645–677.

Saadi, Jana I., Leah Chong, and Maria C. Yang. 2024. The effect of targeting both quantitative and qualitative objectives in generative design tools on the design outcomes. *Research in Engineering Design* 35: 409–425.

Sánchez Pérez, Ana. 2021. *Flatwriter: Representación de una realidad cambiante a través de la obra de Yona Friedman. Trabajo fin de grado*. Universidad Politécnica de Cartagena.

Sánchez-Pérez, Ana, Manuel A. Ródenas-López, Martino Peña Fernandez-Serrano. 2022. Flatwriter today. Towards Friedman's Utopia through generative design. In *Architectural graphics: Volume 3—Graphics for education and thought*, ed. Manuel A. Ródenas-López, José Calvo-López, and Macarena Salcedo-Galera. Springer.

Shea, K., R. Aish, and M. Gourtovaia. 2005. Towards integrated performance-driven generative design tools. *Automation in Construction* 14 (2): 253–264.

Srivastava, Jagriti, and Hiroshi Kawakami. 2023. Systematic review of difference between topology optimization and generative design. *IFAC-PapersOnLine* 56 (2): 6561–6568.

Szabó, Kristóf. 2023. Investigation of the Applicability of Topological Methods. In: *Vehicle and Automotive Engineering 4. Select Proceedings of the 4th VAE2022*, Miskolc, Hungary, ed. Károly Jármai and Ákos Cservenák. Springer.

Tee, Yun Lu, Tobias Maconachie, Philip Pille, Martin Leary, Truong Do, and Phuong Tran. 2021. From nature to additive manufacturing: Biomimicry of porcupine quill. *Materials and Design* 210: 110041.

Teixeira, Joao, Cecília Ogliari Schaefer, Bárbara Rangel, Lino Maia, and Jorge Lino Alves. 2023. A road map to find in 3D printing a new design plasticity for construction—The state of art. *Frontiers of Architectural Research* 12: 337–360.

Ubalde, Santiago Mañé. 2023. A study on the application of generative design in aircraft conceptualization. Trabajo Fin de Máster. Universitat Politècnica de Catalunya.

van Houten, F., R. Wertheim, A. Ayali, E. Poverenov, G. Mechraz, U. Eckert, H. Rentzsch, I. Dani, M. Willocx, and J.R. Duflou. 2021. Bio-based design methodologies for products, processes, machine tools and production systems. *CIRP Journal of Manufacturing Science and Technology* 32: 46–60.

Vardouli, Theodora. 2023. Bioptemes and Mechy Max systems: Topological imaginations of adaptive architecture. In *Disruptive technologies: The convergence of new paradigms in architecture*, ed. Philippe Morel and Henriette Bier. Springer.

Vigil-Fernandez, Carlos, Valentin Gomez-Jauregui, Cristina Manchado, Pedro Lastra-González, and César Otero. 2024. Generative design and analysis of a bicycle frame. In: *Advances in Design Engineering IV Proceedings of the XXXII INGEGRAF International Conference*, 21–23 June, Cádiz, Spain, ed. Cristina Manchado del Val, Miguel Suffo Pino, Ramón Miralbes Buil, Daniel Moreno Sánchez, and Daniel Moreno Nieto.

Vuletic, Tijana, Alex Duffy, Laura Hay, Chris McTeague, Laura Pidgeon, and Madeleine Grealy. 2018. The challenges in computer supported conceptual engineering design. *Computers in Industry* 95: 22–37.

Wadinambiarachchi, Samangi, Ryan M. Kelly, Saumya Pareek, Qiushi Zhou, and Eduardo Velloso. 2021. The effects of generative AI on design fixation and divergent thinking. In *Proceedings of the 2024 CHI Conference on Human Factors in Computing Systems*. arXiv:2403.11164.

Walmsley, Kean. 2022. Cutting-edge practical research on generative design, IoT and digital twins. In: *Innovation in construction. A practical guide to transforming the construction industry*, ed. Eujin Pei and John Roberts. Springer.

Wang, Weijun, Chen Zheng, Feng Tang, and Yicha Zhang. 2021b. A practical redesign method for functional additive manufacturing. *Procedia CIRP* 100: 566–570.

Wang, Hung-Hsiang, and Xiaotian Deng. 2022. The role of fluid intelligence in creativity: The case of bio-inspired design. *Thinking Skills and Creativity* 45: 101059.

Wang, Hui, Wenfeng Du, Yannan Zhao, YingqiWang, Runqi Hao, and Mijia Yang. 2021a. Joints for treelike column structures based on generative design and additive manufacturing. *Journal of Constructional Steel Research* 184: 106794.

Weisberg, Robert W. 2006. *Creativity understanding innovation in problem solving, science, invention, and the arts*. Wiley.

Watson, Marcus, Martin Leary, David Downing, Milan Brandt. 2023. Generative design of space frames for additive manufacturing technology. *The International Journal of Advanced Manufacturing Technology* 127:4619–4639. https://doi.org/10.1007/s00170-023-11691-9

References

Willocx, Mart, Amir Ayali, and Joost R. Duflou. 2021. Reprint of: Where and how to find bio-inspiration? A comparison of search approaches for bio-inspired design. *CIRP Journal of Manufacturing Science and Technology* 34: 171–177.

Yadav, Pratik, Vinay Yadav, Vishal Francis, and Narendra Kumar. 2021. Use of a generative design approach for UAV frame structure optimization and additive manufacturing. In: *Advances in Modelling and Optimization of Manufacturing and Industrial Systems. Select Proceedings of CIMS 2021*, ed. Singh, Ravi Pratap, Mohit Tyagi, R. S. Walia, and J. Paulo Davim.

Yoo, Soyoung, Sunghee Lee, Seongsin Kim, Kwang Hyeon Hwang, Jong Ho Park, and Namwoo Kang. 2021. Integrating deep learning into CAD/CAE system: Generative design and evaluation of 3D conceptual wheel. *Structural and Multidisciplinary Optimization* 64: 2725–2747.

Zhang, Wei, Mark Price, Trevor Robinson, Declan Nolan, Peter Kilpatrick, and Sakil Barbhuiya. 2020. Gene-inspired development of innovative design: Principles and algorithm. *Procedia CIRP* 91: 838–843.

Chapter 7
Concluding Remarks

Abstract The chapter presents the concluding remarks of the book. The book aimed to introduce generative design to both specialized and non-specialized audiences, focusing on its application in aerospace. Generative design is closely linked with new creative design processes and a new manufacturing process. Specifically, the methodology is suited for additive manufacturing. In this sense, it can help obtain novel, lightweight designs that align with sustainable production objectives. Technological advances, such as big data, computing, and artificial intelligence, significantly enhance its capabilities. The suitability of this methodology in aerospace has been discussed by presenting several examples in previous chapters.

The present book aims to introduce generative design to specialized and non-specialized readers, presenting a big picture going from the creative process to manufacturing. Although this design approach has already been successfully applied in various sectors, the book focuses on aerospace.

The design process is closely tied to creativity; thus, this study aims to discuss the concept of creativity and explore its origins. As generative design can be included in the exploration process of manufacturing, it becomes imperative to address both manufacturing and assembly operations. Machining as a mature and intensive process in aerospace and additive manufacturing as a novel and potentially disruptive process are briefly presented. Assembly operations are essential in aerospace, but additive manufacturing can allow for reshaping of assembly operations (i.e., the process can help reduce the number of assembly operations required).

Understanding the context of the current economy is essential. In this sense, sustainable production is one of the critical issues for companies and society. As generative design aims to optimize structures to be more efficient, ideally, the geometry will be more sustainable than other alternatives or former geometries in the case of redesign. However, this intuitive notion requires careful consideration. A proper assessment of the parts in terms of their life cycle can provide designers with suitable metrics for evaluating sustainability.

Advanced design methodologies, such as generative design, take advantage of the significant technological advancements of the last few decades. Thus, advances in computing, big data, and artificial intelligence are key technologies that facilitate methodologies such as generative design.

Generative design evolves prior structural optimization methodologies, such as shape, size, and topology optimization. The objective of these methodologies is explained in the text, but particular emphasis is placed on topology optimization as this methodology is more connected to generative design. Topology optimization aims to explore the design space and develop solutions that help designers create more innovative designs.

Generative design includes many design variables to develop the simulations, requiring a designer's contribution to drive the process. Materials, manufacturing processes, load cases, and constraints are considered to reach an optimized solution. Thus, the approach can be viewed as a comprehensive analysis. However, there is still space for improvement. Stoiber and Kromoser (2021) discussed the "best possible state of a defined problem" concept in an optimization process. In this context, evaluating whether the generative design results can be the best possible is pertinent. Despite being a comprehensive approach, it is clear that factors such as economy, environment, functionality, durability, and aesthetics are not included (i.e., at least, they are not thoroughly addressed in the process).

The advance of big data analytics may help improve generative design capabilities in the future, allowing for the inclusion of more design factors. In any case, the methodology provides numerous results that surpass those of traditional conceptual design. These results aid designers in generating concepts and developing near-ready-to-manufacture designs.

The book presented some examples for aerospace applications, such as the analysis of brackets, drone frames, and the bionic partition. The results help users understand how the tool works. These results are unconventional and innovative, resembling those found in biology.

An essential aspect of generative design is that its outcomes are especially suited for additive manufacturing. This technology enables the manufacture of parts with complex geometries. Thus, it can directly create parts based on the generative design results. Traditional manufacturing techniques can hardly achieve these results.

Generative design is already implemented in several sectors, with major software providers offering commercial solutions that are currently being tested in aerospace applications. Universities are also using these tools to instruct students in this new approach to design. Thus, an increasing trend in the use of this methodology is expected in the coming years.

There is an ongoing debate about the use of in creative tasks. However, industrial design seems less related to the arts than other activities. For instance, many applications can be found in structural parts that, in the end, will be hidden in assemblies. Thus, it is believed that less controversy will exist in the engineering field. In this sense, it is expected that generative design will gain more relevance in the industry, particularly in aerospace, helping to refine existing parts and assemblies, and, ultimately leading aerospace companies to improve operational efficiency.

Reference

Stoiber, Nadine, and Benjamin Kromoser. 2021. Topology optimization in concrete construction: A systematic review on numerical and experimental investigations. *Structural and Multidisciplinary Optimization* 64: 1725–1749.

The manufacturer's authorised representative in the EU is Springer Nature Customer Service Centre GmbH, Europaplatz 3, 69115 Heidelberg, Germany. If you have any concerns regarding our products, please contact ProductSafety@springernature.com

Printed and bound by CPI Group (UK) Ltd, Croydon, CR0 4YY

26/03/2026

02078969-0005